Pitcher Plants of the Americas

Pitcher Plants
of the
Americas

by

Stewart McPherson

The McDonald & Woodward Publishing Company
Blacksburg, Virginia

The McDonald & Woodward Publishing Company
Blacksburg, Virginia, and Granville, Ohio
www.mwpubco.com

Pitcher Plants of the Americas

Copyright © 2007 by Stewart McPherson
All rights reserved. First printing November 2006
Printed in Canada by Friesens, Altona, Manitoba

10 9 8 7 6 5 4 3 2 1
15 14 13 12 11 10 09 08 07 06

Library of Congress Cataloging-in-Publication Data

McPherson, Stewart, 1983–
  Pitcher Plants of the Americas / Stewart McPherson.
    p. cm.
  Includes bibliographical references and index.
  ISBN-13: 978-0-939923-74-8 (pbk. : alk. paper)
  ISBN-10: 0-939923-74-2 (pbk. : alk. paper)
  ISBN-13: 978-0-939923-75-5 (hardcover : alk. paper)
  ISBN-10: 0-939923-75-0 (hardcover : alk. paper)
      1. Pitcher plants—America—Classification. 2. Pitcher plants—
      America—Identification. I. Title.
  QK917.M37 2006
  583'.75091812—dc22
  2006020423

Contents

Dedicated to

Sir David Attenborough

whose inspiring lifeworks

captivated a generation's interest in natural history

and will always fascinate me

Acknowledgements

The process of writing this book was made possible only through the invaluable contributions of my family, many friends and professional colleagues whom I would like to thank here.

The continual help and support of my parents throughout the past five years made this book possible. I could not possibly have undertaken this project without your support and constructive advice which has enhanced every chapter and every page of this work. Thank you for all of your help over the many years!

I would like to express my special thanks to two very good friends and pitcher plant experts, Jim Miller and Andy Smith. Jim, thank you for the knowledge and expertise which you shared so freely with me and for the field trips on which we ventured together! Your experience in the field spans several decades and was of immense help, and your suggestions certainly improved the text of this book, especially the conservation section. Andy, thank you for so willingly offering to dedicate so many hours to produce the illustrations in this book and for sharing your expert growing skills with me over many years. I will always cherish our muddy treks over the purple heath lands of Dorset and your expert knowledge of the natural world! I am sincerely very grateful indeed to both of you for your help and friendship!

I am also extremely grateful to Mike King who, over many years, has shared his expert knowledge of North American pitcher plants with me and countless other enthusiasts. His willingness to share his time and collection, and his efforts as a national collection holder and conservationist, are inspirational. I would also like to sincerely thank my friends Andreas Fleischmann, Brooks Garcia, Fernando Rivadavia, Gert

Highbattle, Andreas Wistuba and Mike Newlin for their assistance throughout this project and for our talks and reflections on pitcher plants. I am also very grateful indeed to Bob Hanrahan for allowing me to photograph the *Sarracenia* on his outstanding wetland preserve.

I would like to express my gratitude to the following libraries, which offered resources to complete research in preparation for this book. The dedicated staffs at Yale University Beinecke Library, University of Wisconsin-Madison Libraries, Royal Botanical Gardens of Kew Library, Heidelberg University Library, Mannheim University Library, Durham University Library and the British Library in London who went out of their way to help me track down rare and elusive texts and publications.

Finally I would like to sincerely thank all at The McDonald & Woodward Publishing Company who have overseen the completion of this book and helped make it become a reality. In particular, I am especially grateful to Jerry McDonald for his dedication and advice throughout the editing process which greatly enhanced the content of *Pitcher Plants of the Americas* and made the final stages of this project a pleasure to undertake.

Pitcher Plants of the Americas

Introduction

Pitcher plants are the largest and most beautiful of the world's carnivorous plants. They are unique within the plant kingdom in that they produce modfied leaves that form hollow, water-containing vessels that are adapted to trapping and digesting animal prey (figures 1 and 2). In most species the leaf traps, the so-called "pitchers," actively attract prey through lures such as nectar, scents and conspicuous colouration. Indeed, these baits are occasionally so successful that the very largest species of pitcher plants catch not only insects but, in some cases, larger prey including small mice and other rodents. Animals that fall into the plants' hollow, water-filled leaves are unable to escape and eventually drown. Slowly, their remains are digested by secreted enzymes and bacterial action to release nutrients that are absorbed directly by the plant. Through carnivory, the pitcher plants have evolved the ability to acquire nutrients that are otherwise not accessible to noncarnivorous plants and consequently they are able to grow in the most barren and inhospitable of habitats, indeed often in areas where noncarnivorous plants are unable to survive. Added to their intriguing carnivorous nature, the flowers and the highly specialized leaves of pitcher plants are often remarkably beautiful and have captivated the fascination and curiosity of people for centuries.

Figure 1 (facing page, top). *Sarracenia* is the most widespread genus of true pitcher plant in the Americas, and *Sarracenia purpurea* is the most widespread species in the genus.

Figure 2 (facing page, bottom). The true pitcher plants of South America are all in the genus *Heliamphora*. Most species, like *Heliamphora pulchella* shown here, have significantly limited ranges, have been studied very little and are poorly understood.

Figure 3. A population of *Darlingtonia californica* growing alongside a lake in southern Oregon.

Seven genera of pitcher plants are distributed across the world. Five occur naturally across the Americas and two are native to Australia, Southeast Asia and a number of islands in the Pacific and Indian oceans. The American pitcher plants belong to two families. The more well known and widely distributed of these is Sarraceniaceae which comprises *Heliamphora*, which is endemic to the central and western parts of the Guiana Highlands of Venezuela, Guyana and northern Brazil (Figure 2); *Sarracenia,* which is found across much of southern Canada and the eastern United States (Figure 1); and *Darlingtonia*, the most geographically restricted of the three, which occurs along and near a short section of the Pacific Coast of the United States (Figure 3). The remaining two genera found in the Americas that contain pitcher plants belong to the family Bromeliaceae and include *Brocchinia*, native to the central and western parts of the Guiana Highlands in Brazil, Colombia, Guyana and Venezuela (Figure 4), and *Catopsis*, which occurs across much of the area from the extreme southern parts of North America across Middle America to the Amazon River in South America (Figure 5). The Sarraceniaceae are considered "true pitcher plants" because they

Figure 4. A population of *Brocchinia reducta* on Mount Roraima, Venezuela.

Figure 5. *Catopsis berteroniana* growing on the branches of cloud forest trees in southern Venezuela.

produce "pitcher" reservoirs formed from a single leaf, whereas the pitcher plant species in *Brocchinia* and *Catopsis* are known as "bromeliad pitcher plants" since they produce reservoirs contained within several snugly fitting leaves that are part of the plant's leaf rosette. Despite these structural differences, the carnivorous adaptation in both families has evolved in parallel and the process by which the true pitcher plants and the bromeliad pitcher plants trap their animal prey is very similar.

I have prepared *Pitcher Plants of the Americas* with the hope that this work will help increase the general awareness of pitcher plants as a group of unusual, fascinating and attractive flowering plants; provide a broad understanding of their biology, ecology and natural history; and allow readers to become aware of their conservation status and the fact that some species, in particular, are in need of strategies and actions to protect and preserve them. My approach to these objectives has been to prepare text that is brief but current and substantive, and presented in accessible language, while also providing a strong visual record of the pitcher plants and their habitats with the use of photographs and technical drawings. This work does not include new taxonomic descriptions and it is not intended to be either a botanical synopsis or a horticultural guide. Rather, it is a substantive summary of the content, biology, ecology, distribution and conservation status of the five genera that incorporate the American pitcher plants and an exploration of the spectacular diversity that exists within this group. Readers interested in pursuing additional information about this group of plants may begin their search by consulting some of the appropriate printed sources listed in the Bibliography, but especially those by D'Amato, Schnell and Slack, or by visiting some of the web sites that now exist on the subject of carnivorous plants and that are mentioned at various places in the last two chapters of this book.

Several specific themes have had fundamental influences on the goals of my work and the methods I have employed to achieve them. Perhaps first among these is the relationship between the taxonomy of the pitcher plants and the great amount of diversity that exists among individual plants within the species populations. The natural diversity of plants and animals is rarely appreciated in detail or understood in its full bewildering complexity at the conventional level of abstract taxonomy. Traditionally a species is treated as a uniform entity; it is discovered, various specimens are preserved and from these it is classified and described. Yet the totality of individuals that make up the entire population of a species at any given time is substantially more diverse and complex than the simplified taxonomic description of a species would suggest. Each species actually represents an extremely complex spectrum of potential natural diversity pulled in various directions and existing on various visible and non-visible levels. The highly evolved leaves of the American pitcher plants demonstrate this concept well in that the leaves in each species exist as a multitude of different phenotypic expressions which differ among themselves in colour, shape, size and many other aspects. Extending both the potential and presentation of phenotypic diversity among pitcher plants even further is the fact that species in both families readily hybridize, producing even greater diversity while challenging the validity of the existing taxonomy. In this work, I clearly consider many of the various discernable physical expressions found among the American pitcher plants, irrespective of whether these variations do or do not represent botanically valid taxa at this time. I believe that it is essential for the future of these species that a comprehensive approach to their biodiversity be taken so that the astounding genetic diversity represented by the pitcher plants be demonstrated, acknowledged, appreciated and, hopefully, preserved for the future.

The taxonomy of the American pitcher plants is also of importance to this work for a number of reasons. While the taxonomy of some members of the group has been in place and relatively stable for centuries or decades, that of other members is relatively recent and potentially dynamic. Whether long-standing or recent, however, there also are clear indications, both verbal and biological, that future taxonomic work on this group likely will result in changes. The majority of current species named in *Heliamphora*, for example, have been discovered or named only within the past few years, and the information for most species of *Heliamphora* is basic and not always reliable. Similarly problematic for the modern taxonomic structure is the fact that some species in *Brocchinia* and *Sarracenia* readily hybridize. At the infraspecific level, it is not clear whether all taxa that have been described actually warrant formal taxonomic designation or, alternatively, if some distinct variants that have not been described deserve formal taxonomic distinction. For example, it is not clear whether the white flowered taxon of *Sarracenia alata* deserves status as a form, a variety or to be recognized as taxonomically distinct at all. Therefore, in this work I will refer to all formally described taxa with the use of their taxonomic names, and to all undescribed phenotypic expressions with the descriptive names used for them in horticulture. Since the terms "variety" and "forms" have taxonomic connotations, I will adopt the term "variant" to refer to all discernable variation that is currently undefined taxonomically and awaiting possible classification.

Understanding the ecology of a species is essential to comprehend, and to facilitate, the ability of that species to survive and reproduce in its environment. The natural habitat of a given species typically represents the environmental conditions to which it has become best adapted and suited over time. Although there are still many large unanswered questions concerning the basic ecology of the pitcher plants, and undoubtedly much of the ecological information and theories which are presented here will be improved upon or changed by future investigations, the information that is available does provide a foundation from which to proceed. The ecological information presented in this work provides a description of the typical environments of a species in its native habitat (Figure 6). This information has intrinsic value for the natural history of the species; it is important for use in evaluating, devising and implementing conservation needs; and it provides the horticulturist with insight for use in efforts to care for and reproduce plants raised under cultivation. Photographs in this work provide some of the first published visual documentation of the South American species of pitcher plants in their natural environments.

While most of the South American pitcher plants are relatively secure in their natural habitats, those in parts of Middle America and most in North America face rather different circumstances. Human activities, notably modifications to the land surface that affect the water table, surface water movement and vegetation cover have caused significant if not seriously threatening changes to the habitat of all but one species of *Sarracenia*, and habitat even in some parts of the range of that species has been noticeably degraded or lost entirely (Figure 7). In fact, concerns about the habitat and population viability for all species in the genus are so acute that all are listed in *The Convention on International Trade in Endangered Species (CITES)* and one species and two subspecies — *S. oreophila, S. rubra* ssp. *alabamensis*, and *S. rubra* ssp. *jonesii* — are included on the *List of Endangered and Threatened Plants* of the United States. *Darlingtonia*, although to date adversely affected less than *Sarracenia*, could rapidly become endangered because of its limited natural range in and near areas of large and growing human population. Lowland

Figure 6. *Brocchinia reducta* in its typical habitat on the summit of Auyan Tepui, Venezuela.

Figure 7. Much of the natural habitat of *Sarracenia* species in Canada and, especially, the United States, has been destroyed or degraded during the past few centuries. Loss of habitat to urban development and transportation facilities, which also alter the water budget of both the affected sites and the watershed of which they are a part, has accelerated in recent decades.

populations of *Catopsis* in North and Middle America face increasing threats from deforestation and related land-use changes. If the North and Middle American pitcher plants are going to be protected in their native habitats, it is first necessary to inform concerned people about the nature and magnitude of the processes that create the threats to the plants and their habitats, then to devise and implement strategies with which to conserve, or re-establish, those places and populations. Similarly, it is necessary to support the continued protection of those pitcher plants that are not presently endangered. The economic needs of communities and nations underlie all conservation attempts, yet with rational, comprehensive approaches to the protection and management of natural resources, pitcher plants and other parts of natural systems can be appreciated and preserved for the benefit and greater well-being of us all. Toward this end, I have provided a summary of the status of the American pitcher plant genera in their natural state along with a review of the major forces that have threatened some populations and species and that have protected, or are being implemented to protect, other populations and species.

Horticulture plays an important and expanding role in creating and maintaining awareness of pitcher plants, and in their conservation. A number of pitcher plants have been raised by horticulturists and some cultivated strains have become standard stock among nurseries and growers. The availability of cultivated specimens is an important step in both increasing the number of living plants representative of some species and in taking some pressure off of the remaining wild populations of those species. Ethical nurseries are important entities in the propagation and distribution of cultivated plants (Figure 8). Consistent with this reasoning, I have identified several cultivated strains of pitcher plants and several nurseries that pioneered the propogation and ethical distribution of pitcher plants, and provided contact information for those growers interested in obtaining plants from responsible nurseries.

I have organized *Pitcher Plants of the Americas* in the following way. After the Introduction, I provide an overview of the taxonomy, world distribution and trapping methods of carnivorous plants of the world, and then I outline the taxonomy, evolution and distribution of the American pitcher plants. Thereafter, I devote one chapter each to the five genera of American pitcher plants, and for each species in each genus of pitcher plant I provide substantive information about its morphology,

Figure 8. Multiple specimens of the various cultivated strains of pitcher plants help to increase awareness of the plants and their conservation needs as well as to maintain viable stocks of living specimens. This population of *Sarracenia* cultivars is part of the collection of Shropshire Sarracenias, an ethical nursery located in Telford, Shropshire, England.

biology, ecology, habitat requirements and distribution. Two additional chapters consider the conservation status of the pitcher plants, with particular attention to the processes that have adversely affected the North American *Sarracenia* taxa and protective strategies that are emerging, and the role of horticulture in developing cultivated varieties of pitcher plants, distributing these cultivars and contributing to the protection and conservation of natural populations of pitcher plants.

Carnivorous Plants

In 1875, Charles Darwin published a book entitled *Insectivorous Plants*. In this work, Darwin presented a series of meticulous experiments by which he conclusively demonstrated that a species of sundew, a small plant he observed in the wetlands of England, possesses the specialized ability to trap, kill and digest insects as a source of nourishment. He concluded that the sundew contravenes the conventional rules of the natural world and is a plant that preys upon animals.

During the 130 years since the appearance of *Insectivorous Plants*, more than 590 species of plants from around the world have been shown to be carnivorous (Figure 9). Each has evolved a method of attracting, trapping and digesting insects and other small animals as a means of augmenting nutrients which enable it to survive in environments that are sometimes relatively barren and deficient in nutrients (Figure 6).

All known carnivorous plants are flowering plants (angiosperms) and are capable of photosynthesis, but nonetheless all obtain nutrients from their carnivorous habits and also to varying degrees through their root systems. In most species of carnivorous plants, the function of the roots is greatly reduced and active uptake of minerals is limited. The carnivorous plants of the world can be divided into five main groups based on the trapping methods which they employ (Table 1).

The largest and most diverse of these groups consists of the 240 species of sticky flypaper plants belonging to the genera *Byblis, Drosera, Drosophyllum, Ibicella, Pinguicula, Roridula* and *Triphyophyllum*, distributed among seven families in three orders (Caryophyllales, Ericales and Lamiales) (Table 1). The sticky flypaper plants produce leaves that are

---

Figure 9 (facing page). *Drosera roraimae*, a sticky carnivorous sundew.

Table 1. An Overview of the Carnivorous Plants of the World

ORDER	FAMILY	GENUS
Caryophyllales	Dioncophyllaceae	*Triphyophyllum*
	Droseraceae	*Aldrovanda* (Water Wheel Plant) *Dionaea* (Venus Fly Trap) *Drosera* (Sundews)
	Drosophyllaceae	*Drosophyllum* (The Dewy Pine)
	Nepenthaceae	*Nepenthes* (Tropical Pitcher Plants)
Ericales	Roridulaceae	*Roridula*
	Sarraceniaceae	*Darlingtonia* (Cobra Lily) *Heliamphora* (Marsh Pitcher Plants) *Sarracenia* (Trumpet Pitcher Plants)
Lamiales	Byblidaceae	*Byblis* (Rainbow Plants)
	Lentibulariaceae	*Genlisea*
		Pinguicula (Butterworts)
		Utricularia (Bladderworts)
	Martyniaceae	*Ibicella* (Devils Claw Plant)
Oxalidales	Cephalotaceae	*Cephalotus* (Australian Pitcher Plants)
Poales	Bromeliaceae	*Brocchinia* *Catopsis*

lined with sparkling droplets of sticky glue that adhere to, overwhelm and eventually suffocate and kill insects of various sizes (Figure 9). The leaves of some species of *Drosera* (Figure 10) and *Pinguicula* (Figure 11)

Number of Species	Distribution	Trapping Method
1	Africa: Ivory Coast, Liberia, Sierra Leone	Sticky Flypaper
1	Africa, Asia, Australia, Europe	Bear trap
1	North America: United States — North Carolina	Bear trap
At least 150	Africa, Asia, Australia, North America, South America	Sticky Flypaper
1	Europe: Portugal, western Spain	Sticky Flypaper
At least 90	Madagascar, southern Asia, northern Australasia	Pitcher Trap
2	Africa: South Africa	Sticky Flypaper
1	North America: United States — Oregon and California	Pitcher Trap
At least 15	South America: Guiana Highlands	Pitcher Trap
8	North America: southern Canada, eastern United States	Pitcher Trap
At least 6	Northwestern Australia	Sticky Flypaper
At least 20	Africa, South America	Cork Screw Trap
At least 75	Asia, Europe, North America, South America	Sticky Flypaper
At least 215	Africa, Asia, Australia, North America, South America	Bladder Trap
1	Middle America	Sticky Flypaper
1	Southwestern Australia	Pitcher Trap
2	South America: Guiana Highlands	Pitcher Trap
1	North, Middle and South America	Pitcher Trap

move relatively quickly and twist and wrap around prey to ensure that it is caught and digested effectively. The traps vary greatly in size from a few millimeters to over 70 cm in length.

Figure 10. The sticky leaves of *Drosera filiformis* glistening at dawn at a site in southern Alabama.

The second group consists of the 220 species belonging exclusively to the genus *Utricularia* (Table 1). These remarkably widespread plants, collectively known as the bladderworts, grow in virtually every country on every continent except Antarctica. They produce small, inconspicuous traps usually buried below ground or under water. The trapping apparatus of these plants consists of a small hollow bladder which maintains a low internal pressure. Trigger hairs surround a small door on the side of the bladder and, when stimulated, cause the door to open. The ensuing rush of air or water caused by the change of internal pressure draws the prey into the bladder where it is retained, killed and digested.

In most cases the traps of the bladderworts are 3–8 mm long and they generally catch larvae and other minute organisms.

The third group includes the most famous carnivorous plant of all — the Venus fly trap, *Dionaea* (Figure 12) — and its lesser-known aquatic relative, *Aldrovanda*. Both genera are in the Droseraceae (Table 1). These two monotypic genera catch their prey in essentially the same manner. They both have leaves which form a small pair of hinged lobes which rapidly snap shut like a bear trap and remain shut until the victim is dead and digested. The traps of *Dionaea* are generally less than 25 mm in length and catch insects of various sizes while those of *Aldrovanda* are less than 7 mm in length and trap small aquatic organisms.

The fourth group consists of the twenty or so species of the genus *Genlisea*, the corkscrew plants (Table 1), which produce long, hollow, twisted fork-shaped traps underground. A narrow slit runs along the length of these traps and allows insects to enter the hollow structure, but thereafter hinders their escape. The corkscrew plants trap mainly minute organisms smaller than 1 mm in diameter.

Figure 11. The sticky leaves of a butterwort, *Pinguicula planifolia*.

The fifth group includes the 115 species of pitcher plants, the largest and most spectacular of all insect-eating plants (Figure 13). The pitcher plants are found in the orders Caryophyllales, Ericales, Oxidales and Poales (Table 1), and are distributed across southeastern Asia, southwestern Australia, islands of the Indian and western Pacific oceans and the Americas. All members of this group produce modified leaves which collect and hold rain water and secrete digestive liquids. Insects and, occasionally, small rodents are attracted to these specialized leaves by secretions of sweet nectar and other baits and, once at the leaves, may fall into the hollow traps where they drown and are digested by secreted enzymes. The pitcher plants are arranged in two subgroups based on their traps.

The first subgroup consists of the true pitcher plants which produce cup shaped or tubular traps in modified individual leaves. The true pitcher plants include the genera *Cephalotus* from the southwest coast of Australia; *Darlingtonia* from the west coast of North America; *Heliamphora* from the Guiana Highlands of South America; *Nepenthes*

Figure 12. The leaves of the notorious Venus fly trap, *Dionaea muscipula*.

Figure 13. *Nepenthes ampularia,* a tropical pitcher plant from the rainforests of Southeast Asia.

from southeastern Asia, Australia and islands of the Indian and western Pacific oceans; and *Sarracenia* from North America (figures 1, 2, 3 and 13). The true pitcher plants produce the largest traps of all carnivorous plants; the pitchers of several species of *Sarracenia* grow to more than 100 cm in height while the stouter traps of various species of *Nepenthes* exceed 1.5 litres in volume and are known to occasionally catch rodents, including rats, in the wild.

The second subgroup of pitcher plants consists of representatives from two genera of tank bromeliads, bromeliads with leaf rosettes that are adapted to collect and store water, including one species of *Catopsis* from the tropical and subtropical regions of the Americas and two species of *Brocchinia* from the Guiana Highlands of Venezuela, Brazil and Guyana (figures 4, 5 and 6). This group differs from the true pitcher plants in that the individual leaves are not hollow, but rather multiple leaves in the leaf rosette fit together snugly to form a water-collecting structure which traps insect prey.

The American Pitcher Plants and Their Evolution

The five genera of American pitcher plants are found in two families, Bromeliaceae and Sarraceniaceae (Table 1). The former family comprises the bromeliads, a diverse group of flowering plants that, with the exception of a single species that uniquely occurs in western Africa, are endemic to tropical and subtropical regions of the American continents. The tank bromeliads are a subgroup of this family and are characterized by strap-shaped foliage that is arranged in a tight, circular rosette which efficiently collects rainwater and organic debris as an alternate means of sustinence. Species in two genera of tank bromeliads (*Brocchinia* and *Catopsis*) have evolved the ability to attract and passively trap insects and are the only bromeliads that are known to possess carnivorous traits. Sarraceniaceae consists of three genera — *Darlingtonia*, *Heliamphora* and *Sarracenia* — which all produce tubular or funnel-shaped, water-collecting leaves that have evolved the ability to attract and catch insects and other small animals (Figure 14). The Sarraceniaceae are distributed across parts of tropical South America and much of North America, primarily in the eastern part of the continent but also in a small part of the western coastal region (Figure 15).

As two groups of plants, the Sarraceniaceae and the Bromeliaceae are not closely related. They are both flowering plants (angiosperms) but their ancestors separated to follow different evolutionary paths 150–200

Figure 14 (facing page). The vibrantly coloured foliage of *Heliamphora chimantensis* is particularly spectacular where a group of plants grow closely together.

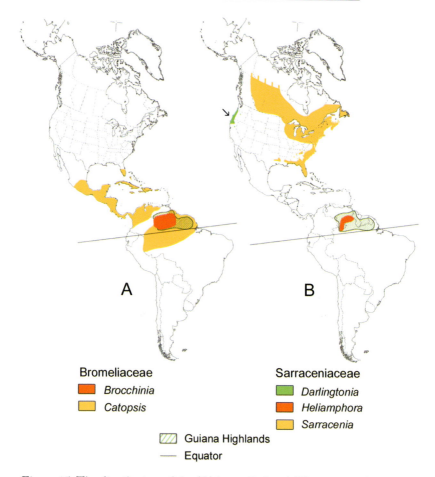

Figure 15. The distribution of the (A) bromeliad and (B) sarraceniad pitcher-plant genera in the Western Hemisphere. Note the Guiana Highlands in northern South America and the small range of the sarraceniad genus *Darlingtonia*, shown by the arrow, on the west coast of North America.

million years ago when the ancestral angiosperms diversified into the monocotyledons (seed producing plants that produce an embryo with a single leaf, including the ancestor of the Bromeliaceae) and the dicotyledons (seed producing plants that produce an embryo with paired leaves, including the ancestor of the Sarraceniaceae).

Unfortunately, virtually no fossil evidence or pollen records exist to suggest where or when the American pitcher plants first evolved; the sequence, chronology or locations at which their various specialized adaptations first came about; or the historic patterns of their distribution and migrations. Interpretations of the evolutionary history of the group are, therefore, based strongly on the accepted relationships between the pitcher plants and some of their relatives with better known evolutionary histories and *a priori* inferences based on their contemporary biology and distribution.

Today, the Bromeliaceae and Sarraceniaceae are confined to the Americas, so it seems likely that all five genera evolved in the Western Hemisphere. This would place the lower limit of their appearance at or after the end of the Cretaceous Period (about 70–65 million years ago).

The evolution of the carnivorous tank bromeliads appears to be less complex than that of the true pitcher plants. The changes which these plants underwent to become carnivorous seem to be small and the required morphological adaptations and changes probably occurred relatively fast. In the cases of both *Brocchinia* and *Catopsis*, only one or two of approximately twenty species in each genus are carnivorous. Tank bromeliads are naturally predisposed to develop carnivorous adaptations. All tank bromeliads inherently possess a rudimentary, water-containing "pitcher" which stores water and traps organic debris, including the occasional entrapment of insects. The evolutionary model of *Brocchinia hechtioides*, *Brocchinia reducta* and *Catopsis berteroniana* simply suggests that all three species evolved first as noncarnivorous tank bromeliads and later developed adaptations that increased the attraction, entrapment and digestion of insects. Important among these adaptations were the ability of the leaves to develop a waxy cuticle, a bright colouration, a coating of reflective white powder, and a vertically enlarged tank structure (Figure 16) — all of which increased the incidence of insect trapping and thereby transformed these species into regular carnivorous plants.

The lack of diversity of carnivorous species in *Brocchinia* and *Catopsis*, and the comparatively small degree of their specialization (for example, they possibly lack the ability to secrete enzymes) suggests that the carnivorous tank bromeliads evolved their carnivorous traits relatively recently. However, it remains unclear whether the carnivorous *Brocchinia* and *Catopsis* represent the first (and only) carnivorous tank bromeliads, or the end of a once more numerous and more varied lineage. Even though the carnivorous *Brocchinia* and the carnivorous *Catopsis* evolved independently, both lineages were clearly influenced by the same

Figure 16. The spectacular golden foliage of *Brocchinia hechtioides,* one of the carnivorous tank bromeliads, helps to define this hillside in southern Venezuela.

evolutionary processes and evolved very much in parallel with one another. The carnivorous species of *Brocchinia* and *Catopsis* share profound similarities and, for the most part, possess comparable adaptations.

The evolution of the Sarraceniaceae is more complex than that of *Brocchinia* and *Catopsis.* According to one current interpretation (Pietropaolo and Pietropaolo, 1986) the evolution of the Sarraceniaceae began when a primitive dicotyledon developed a pubescence on the surface of its leaves, probably as a defence against insect predators or as a means of insulation. At that time, the ancestors of Sarraceniaceae had broad, flat leaves perhaps similar to the foliage of the Actinidiaceae which are believed to be the closest living relatives of the Sarraceniaceae. The pubescence naturally had the ability to retain moisture, and occasionally after rain or dew small gnats would became trapped on the wet, hairy surface of the leaves. Bacteria broke down the small decomposing remains of the inadvertently trapped gnats and released nutrients onto the foliage which was absorbed directly into the leaves. This gave the ancestors of Sarraceniaceae a slight competitive advantage over the surrounding vegetation and in effect drove the evolution of this plant in directions that improved the efficiency of the chance trapping of insects.

As evolution proceeded, the shape of the plants' leaves changed. The centres of broad flat leaves dipped and the leaf margins rose to become more cup-shaped, a trait that allowed the leaf to catch and retain moisture more efficiently. Over millions of years the process continued; the leaves deepened and eventually gave rise to upright, hollow, roughly conical-shaped leaves that collected and retained water, similar to those of the modern Sarraceniaceae. The original pubescences that lined the leaves evolved into the downwards-pointing hairs which cover the interior leaves of all modern Sarraceniaceae. Specific characteristics, such as nectar secreting glands, waxy linings and enzyme secreting glands, slowly emerged as the lineage evolved an increasingly efficient ability to attract, trap and digest insects.

The reason, time and place that the primitive Sarraceniaceae diverged into the modern *Darlingtonia*, *Heliamphora* and *Sarracenia* remains unclear, but at least two evolutionary models are available to explain the evolution of the Sarraceniaceae. One of these is the traditional model (Slack, 1979), and the other presents my views of how the group might have evolved.

According to the traditional model, early Sarraceniaceae closely resembled modern *Heliamphora* and bore simple, short, cup-shaped leaves that filled with rainwater. This *Heliamphora*-type ancestor of the family appears to have been present on the ancestral American continents following their separation from Africa and Europe during the Cretaceous Period.

Some populations of the *Heliamphora*-like ancestor occupied the undissected plateau of the Guiana Highlands. However, as this plateau surface eroded during the Paleogene Period (65–23 million years ago) and formed the modern tepuis and other isolated high summits, populations of the early *Heliamphora*-like ancestor became isolated on the remnant plateaus and slowly diversified into the modern species of *Heliamphora,* although they retained an ancient form. On the North American continent, some populations of the early *Heliamphora*-like ancestor adapted to the changing environment by evolving increasingly tall and tubular leaves that could compete amidst dense marsh grasses of the wetlands of North America, thus giving rise to the tall *Sarracenia/Darlingtonia*-type pitcher plants. Some of these populations in the western coastal region of the continent became isolated from those farther east, probably due primarily to changes in topography — mountain building in the western part of the continent and subsequent drying of the continental interior — and eventually evolved into the genus *Darlingtonia*. The remaining *Heliamphora*-like ancestors diversified to become the modern *Sarracenia* which occur across much of the eastern part of the continent.

The traditional model of Sarraceniaceae evolution considers *Heliamphora* to be the most ancient genus of the family and, indeed, this theory has long been supported by the perceived simplistic cup-shaped leaf structure of most modern species of *Heliamphora*. However, I believe that many inconspicuous morphological traits of *Heliamphora* challenge this traditional interpretation and imply that, while *Heliamphora* is indeed ancient, it is nevertheless the most highly evolved and specialized of all Sarraceniaceae. Three anatomical characteristics of *Heliamphora* support this position.

Firstly, despite an apparent simplistic leaf shape, the complexity of *Heliamphora* traps is greater than that of *Darlingtonia* and *Sarracenia* and, indeed, possibly all other genera of pitcher plants. One characteristic in particular — the unique drainage mechanism of *Heliamphora* leaves — demonstrates this point well. Unlike all other pitcher plants, *Heliamphora* has evolved the ability to directly control the amount of rainwater which is contained within its foliage. All *Heliamphora* species have either a drainage slit or a drainage hole at the midsection of the leaf which enables excess water to overflow from the leaf before the trap fills with rain and topples over due to the excess weight (Figure 17). The water line within the leaf is maintained at a level that always equates to a weight of water which can be supported by the foliage without damage so, unlike the leaves of *Sarracenia,* those of *Heliamphora* never bend

Figure 17. The drainage hole of *Heliamphora* as viewed from inside the leaf. The interlocking, downwards-pointing hairs act as a filter and ensure that trapped prey is retained within the pitcher, even as excess fluid drains away.

over and break in the rain. The drainage system is, however, much more complicated than a simple hole — which would enable the remains of trapped insect prey to be lost along with the overflowing water. It actually consists of a narrow pore that is lined with interlocking downwards-pointing hairs which filter the overflowing water and prevent the loss of trapped insect remains so that, even during heavy rain, trapped insect prey is retained within the leaf. The species of *Heliamphora* which possess a drainage slit rather than a drainage hole function in essentially the same manner. No other genus of pitcher plants has developed such an ingenious method of self-regulation and the benefit of this characteristic to *Heliamphora* species in the ultra-moist environment of the tepuis is assuredly very significant. Although all other genera of true pitcher plants (*Cephalotus, Darlingtonia, Sarracenia, Nepenthes*) have evolved "lids" or "hood"-type structures of various forms, most species of these genera remain nevertheless vulnerable and are easily damaged by becoming over-filled during heavy rains.

The second attribute of *Heliamphora* which I present as an indication of the highly evolved nature of the genus is the flower, which is among the most sophisticated flower of all carnivorous plants in terms of its adaptations to prevent self-pollination. The flowers of *Darlingtonia* and *Sarracenia* are frequently fertilized by their own pollen, whereas the unique structure of the *Heliamphora* flower and the mechanisms of insect-symbiosis on which the release of the pollen depends, strongly suggests a very extensive period of evolution which likely would have required long-term climatic and ecological stability which the Guiana Highlands uniquely offered.

The third trait of *Heliamphora* which suggests a highly evolved morphology is the presence and distribution of downwards-pointing hairs on the interior of the leaves as a characteristic for retaining trapped prey. Several species of *Heliamphora* bear two or three separate and very distinctive types of hairs in addition to a waxy cuticle to trap and retain prey more efficiently than any other genus of pitcher plant.

Additional evidence of an ancient origin of *Heliamphora* comes from examining the distribution of the genus across the Guiana Highlands and relating that to the geological history of the region (Figure 18). The occurrence of *Heliamphora* populations on the summits of tepuis — such as those of *H. neblinae* on Mount Neblina, a tepui that is surrounded by several hundred kilometers of lowland rainforest, inhospitable habitat for the species — suggests very strongly that ancestral populations of *Heliamphora* occupied the plateau before it was dissected. The Guiana Highlands began to erode some seventy million years ago, and as the old surface became fragmented into numerous isolated summits, relict populations of the ancestral *Heliamphora* probably remained on some of those surfaces (Figure 19). These populations then evolved into different species through processes akin to those of island biogeography. This process would explain why a relatively large number of species of *Heliamphora* have evolved within a relatively small area, why most species occur on only one or a few closely spaced summits and surrounding lowlands, why summits with *Heliamphora* populations typically support only one or very few species, and why some summits have no populations of *Heliamphora* at all.

It seems unlikely therefore that the *Heliamphora* species spread to their current ranges after the formation of the isolated summits since *Heliamphora* populate the most inaccessible plateaus of the Guiana

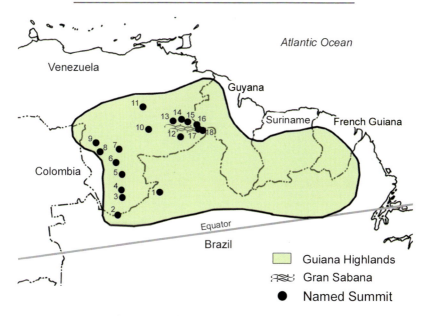

Figure 18. The Guiana Highlands of northern South America and representative summits found within the region. The higher elevations of this ancient but extensively dissected landform are located in its western part.

Highlands, especially those of Bolivar State, Venezuela, yet are absent from larger and more easily accessible tablelands closeby, such as those in the eastern part of Amazonas State, Venezuela. Had *Heliamphora* arrived on the summits of the tepuis relatively recently through processes of wind disposal, then each species might reasonably be expected to occur across more extensive ranges and to overlap more extensively with the ranges of other species.

The preceding observations support a second model of evolution of the Sarraceniaceae, this one postulating a *Darlingtonia / Sarracenia*-type

Figure 19. Chains of waterfalls help to define the vertical nature of the edge of this isolated upland in the heavily eroded Guiana Highlands. Some ancestral populations of pitcher plants likely were present on summits such as this one as the landform eroded, and some of these subsequently diversified into species with highly localized ranges.

plant as the common ancestor of all genera in the family. According to this model, *Sarracenia* diverged from the ancestral form first, followed by *Darlingtonia* and then *Heliamphora*, each lineage evolving separately and in isolation from the others. This theory is supported by the representation of *Darlingtonia* as the intermediate form between *Heliamphora* and *Sarracenia*, especially since it posseses morphological and ecological characteristics that are similar to those of the other two genera but would have been unlikely to have evolved separately and in parallel. The flower of *Darlingtonia*, for example, is similar to that of *Sarracenia* in its simple pollen release mechanism and structure (radially symmetrical five petals and five sepals borne individually on a tall scape), yet it lacks the umbrella-shaped pistil of *Sarracenia* flowers, and its bracts, ovaries and styles are more similar to those of *Heliamphora* than to *Sarracenia*. Likewise, while the distribution of *Darlingtonia* is spatially closer and more similar to that of *Sarracenia* than to *Heliamphora*, the ecology of *Darlingtonia* appears to be more similar to the ecology of *Heliamphora* than to that of *Sarracenia*. *Darlingtonia* and *Heliamphora* share other characteristics too, such as a similar ability to reproduce asexually and form immense clumps that consist of a single genetic strain. Although *Darlingtonia* uniquely reproduces by stolons, it is nevertheless reasonable to consider that such a characteristic might have developed after the separation of *Darlingtonia* and *Heliamphora* and that the dividing trait of *Heliamphora* (most notably as expressed in *H. chimantensis*) is a direct legacy of a characteristic found in the common ancestor of *Darlingtonia* and *Heliamphora* (figures 20 and 21). The fact that *Sarracenia* has a much reduced ability to divide asexually supports the order of divergence suggested above.

An interesting dissimilarity among the three modern genera of Sarraceniaceae is the diversity that exists on differing taxonomic levels. The genus *Sarracenia* consists of eight species with a massive degree of infraspecific diversity whereas *Heliamphora* consists of a much larger number of species with much less infraspecific variation than *Sarracenia*. The genus *Darlingtonia*, however, consists of a single species with relatively little diversity apparent on any taxonomic level. While *Darlingtonia* and *Sarracenia* plants are frequently self-pollinated in their natural habitat, self-fertilization is infrequent in the case of *Heliamphora*

Figure 20. A large clump of *Heliamphora chimantensis* that has developed from a single plant by asexual division.

Figure 21. A large clump of *Darlingtonia californica* that has developed from a single plant by dividing asexually.

due to its complex pollination mechanism. The differing incidence of self-fertilization helps to explain the differences of diversity of each genus. The isolation of *Heliamphora* populations on the highland summits, the relatively small sizes of each population and the frequency of pollination between genetically different individuals within each population helps to explain the relative homogeneity of *Heliamphora* below the species level. The nearly total isolation of *Heliamphora* populations on each summit has driven the development of the comparatively large number of species within the genus. In contrast, the high incidence of self-pollination and the weak mechanisms of reproductive isolation have ensured that *Sarracenia* consists of fewer species than might have developed under different reproductive regimens, but that infraspecific variation within the genus is considerably great. Diversity within *Darlingtonia* is limited by a number of processes and circumstances. *Darlingtonia* occupies the smallest and most environmentally dynamic geographic range of all the Sarraciniaceae. Although the range of the genus is small and the species population is fragmented, the physical isolation of subpopulations is relatively weak and it is reasonable to expect that genetic exchange among these subpopulations could take place easily over comparatively short periods of time. The entire range of the genus probably shifted latitudinally or altitudinally in response to Late Quaternary climatic fluctuations and, if this did happen, earlier expressions of diversity within the genus could have been destroyed and a simplified morphology exhibiting reduced diversity could have emerged from the last glacial refuge to represent the modern genus throughout its current range. The rapid spread of *Sarracenia purpurea* ssp. *purpurea* across glaciated parts of northeastern North America, accompanied by remarkably little morphological diversification of the populations throughout this region, demonstrates that a rapid and genetically conservative dispersal of a genus of Sarraceniaceae is possible. The relative frequency of vegetative reproduction and self-pollination also must contribute to the low diversity found within the genus.

Comparatively little empirical research has examined the evolution and dispersal of the Sarraceniaceae and many significant questions remain, especially concerning distribution. Populations of *S. purpurea* ssp. *purpurea* spread across its vast northern territory are largely homogenous and occupy previously glaciated regions of North America. The territory of this subspecies expanded as the last glacial maximum began to wane some 18,000 years ago, and eventually spread northwards to colonize vast new areas of deglaciated habitat by the time that the current interglacial climate stabilized. Why, therefore, did only *S. purpurea* ssp. *purpurea* spread northwards across a quarter of a continent and inhabit such an extensive area whereas all other *Sarracenia* taxa remained confined to small ranges — albeit perhaps recently established ranges that formed as populations shifted to higher latitudes or altitudes as rising sea levels accompanied the warming interglacial climate? Considering the diversity of distribution vectors which act on the seed of *Heliamphora* (especially wind and water dispersal processes), why has each species remained confined to its ancient, fragmented and environmentally stable range? Why does *Darlingtonia*, which occupies the smallest and potentially most dynamic range, possess such low diversity — especially considering its possible evolutionary relationship to two other genera which have demonstrably high potential to both diversify and be conservative? Considering that *S. purpurea* ssp. *purpurea* occupies such a vast area of North America, why therefore is it so homogenous and why has variation comparable to that seen in the remaining *Sarracenia* species not emerged?

Brocchinia

The name *Brocchinia* honours Italian botanist and director of the Brescia Botanic Gardens, Giovanni Battista Brocchi (1772–1826). Approximately twenty species of *Brocchinia* have been identified across tropical regions of northern South America and the majority occur exclusively within the Guiana Highlands and surrounding areas of Brazil, Colombia, Guyana and Venezuela (Figure 22). All species of *Brocchinia* are relatively large terrestrial or, occasionally, epiphytic tank bromeliads that are adapted to tropical highland and lowland conditions, especially habitats where water is abundant. The genus includes several giant bromeliad species, including the noncarnivorous *B. tatei* which produces expansive foliage up to 1.5 m in diameter. All *Brocchinia* species respond dramatically to differing environmental conditions, and as a result ecophenes (environmental forms) are widespread in all species. The taxonomic classification of species in this genus has been complicated on many occasions by the repeated re-naming of several taxa, resulting in numerous synonyms. Due to the remoteness and inaccessibility of their habitat, relatively little research has been conducted on the general ecology and diversity of *Brocchinia* and, as a consequence, this genus remains among the least known of all bromeliad genera. At this time, several recently discovered bromeliads await study and probable classification.

Two species of *Brocchinia* (*B. hechtioides* and *B. reducta*) habitually trap insects, particularly ants and winged insects, and have been suspected of carnivory for several decades. In the absence of conclusive research it remains unclear whether these bromeliads secrete enzymes of

---

Figure 22 (facing page). *Brocchinia reducta* growing on Mount Roraima, Venezuela.

their own and actively digest trapped prey or, indeed, whether the digestion process relies entirely on secondary organisms such as bacteria, insect larvae and frogs to break down and release as waste some of the
nutrients they obtain from trapped insects. At least one other currently
undescribed species in *Brocchinia* similarly traps insects and may eventually be considered to be carnivorous alongside *B. hechtioides* and *B. reducta*.

The taxonomic classification of *B. hechtioides* and *B. reducta* remains
problematic and it is unclear whether these two taxa represent legiti-
mate separate species or extremes of a single highly variable species.
The only tangible difference between *B. hechtioides* and *B. reducta* is the
structure of the scape, but the general shape and size of the leaf rosette
also differs in a very general way; in *B. reducta* the foliage generally forms
a small, narrow and tightly tubular rosette while the rosette of *B. hechtioides*
is larger and more loosely arranged. The taxonomic problem in defining
the two species lies in the fact that the structure of the leaf rosette varies
greatly in both, resulting in expressions of this characteristic that overlap among individuals from either species to the extent that a continuous range of variation exists between their extremes. It is, therefore, extremely difficult to positively identify specimens that are close to the
boundaries of the two species, a situation that is made even more difficult by the fact that *B. hechtioides* and *B. reducta* are known to readily
hybridize in the wild.

Plant Structure

The two carnivorous species *B. hechtioides* and *B. reducta* are short-
stemmed terrestrial tank bromeliads (Figure 22). The leaves are linear,
20–50 cm long and bright golden yellow to pale green in colour and are
lightly coated with a waxy white powder that is especially apparent on
the interior side of the leaves. The foliage forms a tight upright tubular
rosette which collects and stores up to 600 ml of rainwater. The lower
parts of the leaves are lined with modified hair-like trichomes which
enable absorption of water and nutrients directly into the leaves. Aged
specimens of both species grow on short woody upright or decumbent
stems that grow up to 20 cm in length. Narrrow brown roots emanate
from the base of the plant and can anchor it to virtually bare rock. Between 20–200 individual flowers are borne on the erect 30–80-cm-tall
scapes (Figure 23). The individual flowers of the two species are virtually identical and have long been thought to be dioecious (with male and

Figure 23. The male flower of *Brocchinia reducta*.

female flowers occurring on separate plants). However, recent observations suggest that *B. reducta* and *B. hechtioides* can also occasionally be monoecious (with male and female flowers occurring on the same plant). (My personal field observations have led me to conclude that these plants are overwhelmingly dioecious, but I have observed a number of flowers that appeared to have reproductive organs of both sexes. Further research is certainly required on the structure of the flower of *Brocchinia*). The scape is 2-pinnate in *B. reducta* and 3-pinnate in *B. hechtioides*. The inflorescence usually consists of 1–10 lateral branches (figures 24 and 25). Each flower consists of three white, 3–7-mm-long petals and three yellowish green, 3–5-mm-long sepals. A variable number of stamen (usually 6–8) are present in each series. The anthers are oblong, 2–3 mm long and yellow in colour. The individual flowers open in succession over the course of several weeks, starting with flowers towards the base of the inflorescence. Each flower remains open for 1–4 days. The fruit is a small, 8–25-mm-long ellipsoid capsule which contains 5–70 caudate seeds that are distributed mainly by the wind. The leaf rosette dies shortly after seed is produced, but offshoots typically are produced. In Venezuela, both species flower primarily between November and January, although a minority of both species flower sporadically throughout the year.

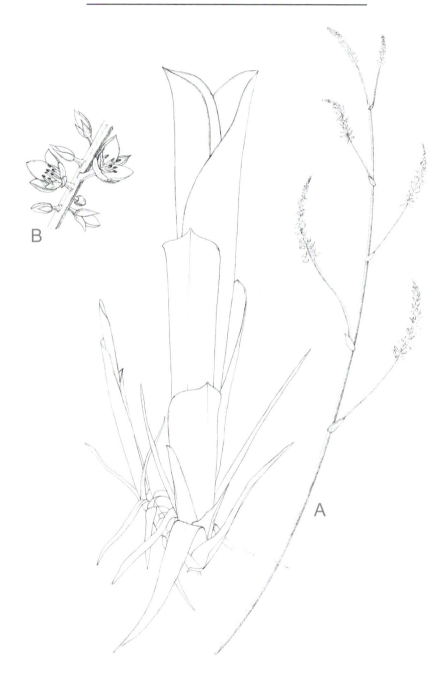

Figure 24. A simplified *Brocchinia* plant, with (A) scape and (B) flowers also shown.

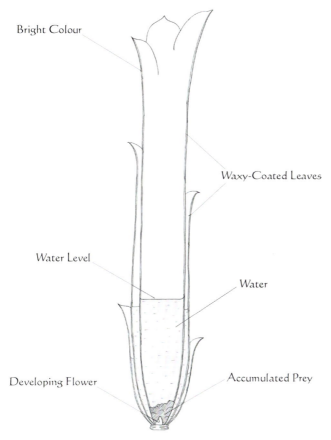

Bright Colour

Waxy-Coated Leaves

Water Level

Water

Developing Flower

Accumulated Prey

Figure 25. The anatomical structure of the above-ground parts of a simplified *Brocchinia* plant, with pitcher contents added.

Trapping Process

The method of carnivory in *B. hechtioides* and *B. reducta* remains poorly understood. According to several field botanists, chemical mimicry is employed by these species as a means of attracting insects. The plants secrete a sweet smelling substance into the liquid contained within the leaf rosettes and this creates an odour that resembles nectar and encourages potential prey to enter the leaf rosettes where, in turn, many fall into the trap. The vivid colouration of the leaves (Figure 26) and the coating of fine white powder, which possibly has ultraviolet reflective properties, are also attractive to insects. Once lured to a plant, insects explore the entrance of the trap and, enticed by the presence of water, the apparent scent of nectar and the remains of previously trapped insects within the depths of the leaves, then venture into the tubular leaf rosette (Figure 27) and climb down the interior of the trap. The new victim finds that, since the leaves are extremely waxy and lined with crumbly white powder, a secure foothold is extremely difficult to maintain. Eventually, the insect falters and slips straight down into the rainwater contained within the rosette or the leaf axils where, unable to scale the slippery interior of the rosette and climb to safety, it drowns or dies of exhaustion. *B. hechtioides* and *B. reducta* have been proven to absorb amino acids more efficiently than noncarnivorous tank bromeliads and in cultivation grow much more vigorously when conditions enable the trapping of insects. The lowland *Brocchinia* populations catch many more insects than the highland populations due to the comparative abundance of insects in the lowlands.

Although *B. hechtioides* and *B. reducta* efficiently trap insects, particularly ants and winged insects, the status of these two species as true carnivorous plants is uncertain since it remains unclear whether they actively secrete enzymes or, indeed, whether the digestive process relies entirely on wastes from secondary organisms such as bacteria, fungi, insect larvae and frogs.

Distribution

B. hechtioides and *B. reducta* are distributed across the Guiana Highlands from western Guyana across southern Venezuela (Amazonas and Bolivar states) to eastern Colombia and the extreme north of Brazil (Figure 28). Both species occur on the summits of the tepuis and in the low lying marshlands across the Guiana Highlands region. *B. reducta* occurs predominantly in western Guyana and Bolivar State, Venezuela, and is particularily common throughout the highlands and lowlands of the Gran Sabana region of Bolivar State, Venezuela. *B. reducta* appears to be uncommon or possibly absent from Amazonas State, Venezuela, eastern Colombia and the extreme north of Brazil. *B. hechtioides* occurs throughout Amazonas State, Venezuela, the extreme east of Colombia

---

Figure 26 (facing page). A mixed population of *Brocchinia hechtioides, Brocchinia reducta* and hybrids.

Figure 27. The tubular leaf rosette of *Brocchinia reducta* collectively forms a "pitcher" trap.

and the north of Amazonas State, Brazil; however, it is most prevalent throughout highland regions between Cerro Avispa and Auyan Tepui in the south of Venezuela. *B. hechtioides* is infrequent but not absent across the Gran Sabana region of Guyana and Bolivar State, Venezuela. The two species occur together most often in the lowlands of Bolivar State, Venezuela, where hybrids have been most frequently observed.

Habitats

B. hechtioides and *B. reducta* predominantly occur on mountain summits across the Guiana Highlands between 1800–2800 m above sea level. At this altitude, the maximum temperature rarely exceeds 26°C and the minimum temperature, even on the highest of the tepuis, is usually above

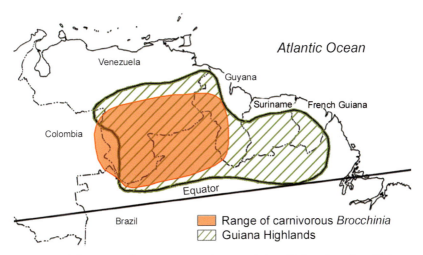

Figure 28. The range of *Brocchinia hechtioides* and *Brocchinia reducta*. Insufficient data exist at this time to accurately define and differentiate the territories of the two species.

4°C during the nights and never falls below freezing. Rainfall is high throughout the year and the very wettest tepuis receive up to 9000 mm of precipitation annually. The excess of rainwater scours the landscape of sediment and renders the summits of many tepuis desolate landscapes of bare rock. This harsh environment usually supports only a small array of hardy plants, of which *B. hechtioides* and *B. reducta* are often among the most prevalent. The carnivorous *Brocchinia* grow successfully in adverse conditions because of the significant advantages they have over most other competing plants; they can procure nutrients and store water by means of their modified leaf rosettes and consequently have evolved buffers against the limitations of their surroundings. Both species can grow on virtually bare rock and require only a small crack or fracture to obtain a roothold (Figure 29). Often the most barren of landscapes support the densest populations of *B. hechtioides* and *B. reducta* as they slowly flourish in the absence of other plants.

In the more hospitable areas of the tepui summits, the two *Brocchinia* species occur in marshy vegetation, often growing in the company of the highland species *Stegolepis guianensis* and *Orectanthe sceptrum*. Both *Brocchinia* species compete efficiently with small plants, but they are generally absent from areas where the surrounding vegetation is taller than 100 cm. Both species display a clear dislike of shaded conditions

Figure 29. The remarkably adaptable *Brocchinia reducta* growing while rooted directly to bare rock.

and grow most vigorously in direct sunlight. *B. hechtioides* and *B. reducta* grow in a variety of habitats, from standing water to well drained, but permanently moist, substrate. They usually grow in acidic peaty sediments, but often grow in little more than a trace of sand or gravel. *B. reducta* has a tendency to occur more frequently in more barren, rocky habitats while *B. hechtioides* is usually more common in more extensively vegetated, hospitable areas. When growing in barren conditions, *B. reducta* readily forms dense island-like clumps.

    B. hechtioides and *B. reducta* also grow frequently in the wetter parts of the lowland savannahs of Venezuela, Guyana and northern Brazil (figures 30 and 31). These lowland populations are usually completely isolated from the highland populations by belts of dense cloud forest which cloak the flanks of the tepuis. Probably very little genetic exchange takes place between the highland and lowland populations since they may be separated from each other by several kilometers and many hundreds of meters of altitude. The lowland populations usually occur between 800–1500 m above sea level. Rainfall is unequal throughout the year and the lowland savannah is frequently prone to drought during the dry season.

Figure 30. A lowland population of *Brocchinia reducta* growing in the Gran Sabana. In warmer conditions, such as is shown here, the leaf rosette of the species is less tightly bound than in plants growing in the cooler, upland environments.

Consequently, *B. hechtioides* and *B. reducta* usually occur in the lowlands only where the soil is permanently wet, mainly at seepage sites and on marshy flood plains. The temperature in the lowlands is several degrees warmer than on the tepui summits; the maximum daily temperature often exceeds 28°C during the dry season while the nights rarely fall below 15°C. Rainfall and humidity are considerably lower than on the tepui summits and generally between 800–2500 mm of precipitation falls annually.

In response to the drier, hotter conditions of the lowlands, *B. hechtioides* and *B. reducta* growing there are usually smaller and have fewer leaves than do the plants on the tepuis. The reduced surface area moderates transpiration and compensates for the more limited availability of water. Water storage in the leaf rosettes is probably very important for the lowland plants and the tubular rosettes of the lowland populations

Figure 31 (following pages). A golden landscape of *Brocchinia hechtioides* in the Guiana Highlands.

are usually less tightly arranged, a trait that facilitates the catching and storing of rainwater.

In the lowlands, *B. hechtioides* and *B. reducta* typically grow amidst short savannah grasses and small shrubs. Both species are generally absent in dense vegetation over 100 cm in height or with 40% or more shade. It is in the lowlands where the two species occasionally occur together and hybridize. Morphologically, the hybrids bridge the two species and create a continuous range of variation, and individuals of either species as well as their hybrids are almost impossible to distinguish in the field with certainty.

General Ecology

B. hechtioides and *B. reducta* have evolved close relationships with a variety of organisms. In a manner very similar to the erect species of *Sarracenia*, various frogs perch within the upper parts of the leaf rosettes and await the arrival of visiting insects. When an insect enters the tubular leaf rosettes, the frog eats it before it is trapped by the plant. Although the frogs effectively rob the *Brocchinia* of its prey, the relationship does have benefits for the plant too. The frog breaks down the insects more efficiently than the plant can and releases the remains as simple nutrients in its waste products which the plant can easily absorb. Of the many frog species that visit the *Brocchinia*, one is entirely transparent and known in Venezuela as *la rana de cristal* (the crystal frog) and *la rana de vidrio* (the glass frog). A multitude of mosquito larvae and small crustaceans inhabit the acidic water contained within the rosettes, and ants, particularly in the lowlands, colonize the dead leaves and make larvae nurseries in the empty leaf axils. *Utricularia humboldtii*, a carnivorous bladderwort, occasionally grows and traps minute aquatic prey in the water-filled axil reservoirs of *B. hechtioides* and *B. reducta* (although much more often it grows in the considerably larger noncarnivorous *B. tatei*).

Once *B. hechtioides* and *B. reducta* mature and flower, the leaf rosette soon dies. In some cases this means the total death of the plant, but usually a side shoot, known as a "pup," develops from the remains of the original plant (Figure 32). The pup grows rapidly and develops an independent root system before the mother plant either dies back or dies altogether. By dividing in this manner, *B. hechtioides* and *B. reducta* can establish large clumps consisting of dozens of individual plants. In some cases *B. hechtioides* and *B. reducta* reproduce by producing long runners

Figure 32. A *Brocchinia reducta* plant which has flowered, produced a "pup" offshoot, and subsequently died back.

that grow into mature plants where they touch the ground (Figure 33). This behaviour, however, has been observed only on a handful of occasions in the wild.

In cultivation, the natural colouration and tubular shape of the leaf rosettes are rarely maintained. Horticulturists usually grow *B. hechtioides* and *B. reducta* in shade, which normal bromeliads prefer, but this causes the leaves to be etiolated and dull green in colour and the rosette to be loosely arranged. Extremely bright light, cool temperatures and acidic soil are required to support healthy natural growth and colouration in these species.

The taxonomic relationship between *B. hechtioides* and *B. reducta* remains unclear. The morphological boundary between the two species is difficult to define and the only significant tangible difference is that *B. reducta* produces a 2-pinnate scape while *B. hechtioides* produces a

Figure 33. An offshoot of *Brocchinia hechtioides* growing on a long runner.

3-pinnate one. Much more generally, the shape and size of the leaf rosette varies between the two species; in *B. reducta* the rosette is generally
small, narrow and tightly tubular while that of *B. hechtioides* is larger and
more loosely arranged. The structure of the leaf rosettes varies greatly
and overlaps between the two species so that a continuous spectrum
exists between the two extremes. It is extremely difficult to positively
identify specimens close to the boundaries of the two species, especially
since *B. reducta* and *B. hechtioides* are known to readily hybridize in the
wild. Consequently it is hard to suggest whether these two taxa deserve
reclassification as one species or not. Although the physiological differences between the two taxa can indeed be significant, ultimately the
shape and proportions of the foliage and flowers are extremely similar.
Several other closely related, currently unnamed taxa of *Brocchinia* display very similar characteristics to *B. hechtioides* and *B. reducta* and may

also be proven to be carnivorous. One presently undescribed candidate is considerably taller than *B. hechtioides* — almost twice the size, although similar in profile — and has black, voluminous leaf axils.

Species Descriptions

Brocchinia hechtioides Mez

Original description: Mez, C. C., 1913, *Repertorium Specierum Novarum Regni Vegetabilis* 12: 414.

The specific epithet *hechtioides* is derived from the similarity in appearance between this species and the genus *Hechtia*, a different genus of bromeliads native to Central America, Mexico and Texas.

The leaves of *B. hechtioides* are 30–60 cm long, 3–6 cm wide and bright yellow or yellowish green in colour (Figure 34). The younger leaves are lighter and greener and have a more prominent coating of white powder than the older leaves. In shape, the leaves are linear and narrow slightly towards the apex. The end of the leaf terminates bluntly, but a small 2–4-mm-long point-like tip is usually present. The mature leaf rosette consists of 5–15 leaves and is 10–30 cm in diameter. The rosette forms an upright tube but the leaves are loosely arranged and most of the water is contained within the leaf axils rather than in the centre of the rosette. The rosette of *B. hechtioides* is much more voluminous than that of *B. reducta*. *B. hechtioides* grow upright and readily forms a short woody stem. The scape is 3-pinnate.

Brocchinia reducta Baker

Original description: Baker, J. G., 1882, *Journal of Botany, British and Foreign* 20: 331.

The specific epithet *reducta* is derived from the Latin *reduco* (reduced or drawn back) and refers to the small size and simplistic tubular shape of the leaf rosette. *B. reducta* is one of the smallest species of *Brocchinia*.

The leaves of *B. reducta* are 20–40 cm long and 2–5 cm wide and golden yellow or light green in colour (Figure 35). The leaf is linear and narrows slightly towards the apex; it also terminates in a small 2–4-mm-long tip-like point. The mature rosette of *B. reducta* consists of 3–10 leaves, which overlap each other and form a water-tight, upright, compact,

Figure 34. A small cluster of *Brocchinia hechtioides*.

tube that is 3–7 cm in diameter. Typically the rosette is one-quarter-full with water, virtually all of which is held in the centre of the rosette. *B. reducta* tends to grow horizontally and crawls along the ground on the length of its stem. The flower is 2-pinnate.

No taxonomic forms or varieties of either *B. hechtiodes* or *B. reducta* have been described. Locally, most populations of both species are very

Figure 35. A typical *Brocchinia reducta* plant.

uniform, and show little diversity, although the shape, size and colour of the leaf rosettes of both species vary substantially among populations on different tepuis and in different lowland areas. The overall variation observed within *B. hechtioides* and *B. reducta* bridges the differences between the two species and makes identification of specimens in the field difficult.

Catopsis

The name *Catopsis* is derived from the Greek words *kata* (hanging) and *opsis* (appearance) and refers to the predominantly epiphytic growing habit of all species of this genus. *Catopsis* are small, brightly coloured tank bromeliads that mostly grow perched on the branches of short, open-canopy trees in tropical and subtropical regions of the Americas. Twenty-one species of *Catopsis* have been described, with the greatest geographic concentration of species occurring between southern Mexico and southern Colombia. The genus is relatively diverse and the shape, size and structure of the foliage vary considerably among the known species. Several specimens have been discovered that do not fit the descriptions of any of the currently defined species and may represent new species. A coating of fine white powder is common to all members of this genus, although this trait is very much reduced in some taxa. *Catopsis* is closely related to the diverse air-plant genus *Tillandsia*. *Catopsis* species, collectively known as the strap air plants, display specialized adaptations towards epiphytic ecology similar to those found in *Tillandsia*, in particular with regards to procuring and storing water and nutrients. Relatively little research has focused on *Catopsis*, and the evolutionary relationship of the genus to other bromeliad species remains unclear.

Only one species of *Catopsis*, *C. berteroniana*, habitually traps insects and is suspected of being carnivorous (Figure 36). *C. berteroniana* has been shown to trap twenty times the amount of insects caught by comparable noncarnivorous tank bromeliads and to derive the majority of nutrients which it requires from trapped prey. It remains unclear whether enzymes are actively secreted by *C. berteroniana* or, indeed, whether secondary organisms are relied upon to digest trapped prey.

Figure 36 (facing page). The elegant foliage of *Catopsis berteroniana*.

Species Description

Catopsis berteroniana (Schultes and Schultes) Mez

Original description: Mez, C. C., 1896, *Monographiae Phanerogamarum* 9: 621.

The specific epithet *berteroniana* honours the Italian botanist C. G. Bertero (1789-1831). In many Central American countries, Colombia and Venezuela, *C. berteroniana* is called *lampera de la selva* (jungle lantern) in reference to its bright colouration.

Plant Structure

C. berteroniana is a short-stemmed, epiphytic tank bromeliad. The leaves are 15–30 cm in length, bright yellowish green in colour and coated with a conspicuous white powder which plays an important role in the trapping process. The leaf blade is lanceolate and parallel veined and the apex is acute. The foliage is arranged in a compact, upright rosette that is adapted to collecting and storing rainwater. The centre of the rosette and the leaf axils form hollow, watertight chambers which contain up to 400 ml of water. *C. berteroniana* grows attached to the branches of trees that grow up to 30-m tall. All of the water and nutrients which *C. berteroniana* requires for growth are collected primarily through the leaf rosettes via rainwater, leaf debris and carnivory, and to a lesser degree through the thread-like brown roots, although these are reduced in function and mainly anchor the plant to the host tree. Aged specimens of *C. berteroniana* form short woody stems up to 5 cm in length. The colouration of *C. berteroniana* is particularly vibrant, and the coating of reflective white powder creates an almost glowing appearance, especially during the twilight hours. Small trees often support as many as twenty bright yellow "lantern-like" *C. berteroniana* plants.

The flowers of *C. berteroniana* are borne on an erect 20–60-cm-tall scape which bears 2–8 lateral branches, each supporting 5–30 individual flowers (Figure 37). Each flower is small and inconspicuous, bracts are 5–8 mm long and oval in shape, the sepals are yellowish green and 0.8–1.4 mm in length (figures 38 and 39). The petals spread slightly and are yellowish white in colour and oval in shape. Similar to *Brocchinia*, the flowers of *C. berteroniana* have long been documented as being dioecious but recent observations suggest that indeed the species may occasionally be monoecious too. Other species of *Catopsis* are known to vary among populations in terms of flower sexuality and this appears to be the case with *C. berteroniana* too. Flowers are produced throughout the year. Fertilized flowers produce an 8–16-mm-long ellipsoid fruit capsule which contains 10–100 caudate seeds which are similar in appearance to the seeds of *Tillandsia* species and are distributed by the wind. The leaf rosette dies shortly after seed is produced, but it readily generates viable offshoots and, as a result, an individual plant usually flowers more than once.

Figure 37. A *Catopsis berteroniana* plant in flower.

Figure 38. A simplified *Catopsis berteroniana* plant, with (A) scape, (B) flower buds and (C) seeds also shown.

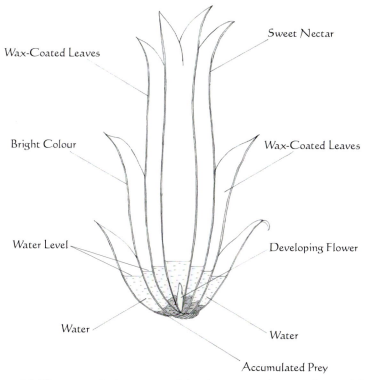

Wax-Coated Leaves

Sweet Nectar

Bright Colour

Wax-Coated Leaves

Water Level

Developing Flower

Water

Water

Accumulated Prey

Figure 39. The anatomical structure of the above-ground parts of a simplified *Catopsis berteroniana* plant, with pitcher contents added.

Trapping Process

In the 1970s and 1980s, various ecologists noted the frequency at which insects were trapped in the leaf rosettes of *C. berteroniana* in comparison to other tank bromeliads and, as a result, it was suggested that the species might be carnivorous. While *C. berteroniana* catches up to twenty times more insects than similar noncarnivorous bromeliad species, many questions remain regarding the classification of *C. berteroniana* as a true carnivorous plant. In particular, the process of digestion remains poorly understood and it remains unclear whether *C. berteroniana* secretes enzymes directly or whether it relies on secondary organisms, mainly insect larvae, to break down the remains of trapped insects. Much like *Brocchinia*, while *C. berteroniana* is not traditionally considered a pitcher plant, it does catch insect prey through a process that is fundamentally identical to that of the true pitcher plants, the only difference being that the plant's trap consists of the entire leaf rosette rather than each individual leaf.

The leaves of *C. berteroniana* create a complex visual illusion to deceive and lure winged insects. All parts of the foliage are lined with a fine, waxy, white powder that efficiently reflects ultraviolet and short-wave light in a manner similar to many tropical flowers. In the ultraviolet-sensitive vision of insects, this makes the leaf rosettes stand out vividly from the background undergrowth and perhaps resemble large yellow blooms. Expecting to find nectar, insects are drawn to the illusion and venture into the centre of the rosette in search of nectarines. However the visiting insect soon finds that a foothold is very difficult to maintain on the waxy, slippery leaves and eventually it falters and falls directly down into the water-filled wells of the leaf axils and drowns (Figure 40). Various organisms, especially mosquito larvae, inhabit the

Figure 40. An ant on the edge of the trap of a *Catopsis berteroniana* plant.

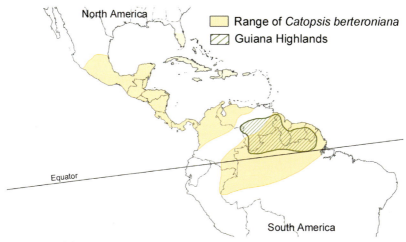

Figure 41. The range of *Catopsis berteroniana*.

wells of *C. berteroniana* and consume the remains of trapped insects. Regardless of its eventual classification as carnivorous or not, *C. berteroniana* does trap insects efficiently. While studying a specimen in the Gran Sabana, Venezuela, over a period of two hours I observed three ants fall into the well of a single plant; the ants were probably attracted by the scent of previously caught prey rather than the visual illusion.

Distribution

C. berteroniana is distributed across most of the American tropics and subtropics (Figure 41). It occurs in southern Florida, across central and southern Mexico, throughout Middle America, and into South America across northern Colombia and northwestern Venezuela. It is absent across the Colombian and Venezuelan Andes but reappears across southeastern Venezuela, the northern parts of the Brazilian Amazon region, eastern Colombia and Peru and throughout Guyana, Suriname and French Guiana.

Habitats

C. berteroniana grows mainly on the branches of trees between 2–6 m in height, but it also occurs at the canopy level of much larger trees that grow up to 30 m in height (Figure 42). It displays a clear preference for bright conditions and grows best on the branches of tree species which have sparse foliage and thereby expose the plants to virtually full

Figure 42. Young *Catopsis berteroniana* plants growing on the branches of stunted cloud forest trees in southern Venezuela.

sunlight. *C. berteroniana* grows poorly in conditions of 35% or more shade, and in 45% shade the leaves are dull green and etiolated. *C. berteroniana* never grows terrestrially; if the branch on which it is growing breaks and falls to the ground, the *C. berteroniana* plant soon dies. It has been reported that *C. berteroniana* grows on telephone wires similar to many species of *Tillandsia*.

    C. berteroniana occurs between 0–2000 m above sea level in a variety of forest types that range from coastal mangroves, to tropical lowland forests, to lower cloud forests. In the tropical lowlands, the climate is consistently warm throughout the year — temperatures are regularly above 28°C during the day and seldom below 15°C at night. Rainfall is often variable, and equatorial vegetation is often subject to periods of drought but *C. berteroniana* overcomes short dry periods by means of the water which it stores in its leaf rosette (Figure 43). It is most prevalent in the more elevated parts of its range, 1500–2000 m above sea level, on the branches of the stunted trees of the lower cloud forest where it grows in the abundant moisture and the cool temperatures that are typically below 25°C during the day and above 9°C during the night.

General Ecology

The seeds of *C. berteroniana* are distributed by the wind. Each seed is attached to many fluffy white 20–30-mm-long filaments which catch the wind and carry the seed even in the faintest breezes. Each inflorescence usually produces several hundred seeds, most of which drift to the ground and never germinate. Some seeds, however, float through the air until their filaments snag the bark or the branches of a tree. Once in place, the filaments absorb rain and dew and provide the seed with moisture to grow. Within the first weeks the seedling grows small leaves and roots and fastens itself to the bark of its host tree. In this early stage, most seedlings die from drought or are washed or blown from the branches, but a minority remain in place and grow to maturity in 2–5 years. After flowering, the leaf rosette soon dies and, although this may

Figure 43. The water-filled leaf rosette of a *Catopsis berteroniana* plant.

mean the total death of the plant, more usually a side shoot, a pup, develops (Figure 44) and grows from the dying remains of the original plant in a manner similar to what happens with *Brocchinia* (Figure 32). The pup grows much faster than seedlings and matures in 2–3 years, and once fully grown it may flower, generate another offshoot and then die back, thereby repeating the process. The leaves of the mother plant fall away as the pup grows, but the dead remains of the root system stay attached to the host tree, so it is easy to count how many times a particular individual has flowered and re-sprouted. In some cases a single plant may resprout ten times, each time moving slightly along the branch as it regrows, thus the life span of an individual can extend to several decades in the wild (Figure 45). Although *C. berteroniana* may flower as soon as it reaches maturity, some specimens do not flower for several years and eventually develop woody stems up to 3 cm long.

Figure 44. A *Catopsis berteroniana* plant which flowered, produced a "pup" offshoot, and then died back.

Figure 45. The old, dead root systems of a *Catopsis berteroniana* plant show that this individual has flowered, produced pups, and died back six times.

As the foliage of *C. berteroniana* is produced, the older leaves turn brown, die back and fall to the ground. This prevents the dull, aging leaves from covering the vividly coloured younger leaves and detracting from the vibrant colouration which advertises the trap.

No taxonomic forms or varieties of *C. berteroniana* have been described and little morphological diversity is apparent across the whole range. In general, individuals are smaller in the lowlands than in the highlands, probably as a response to the hotter, drier lowland conditions.

Darlingtonia

The genus *Darlingtonia* was named by John Torrey in 1853 in honour of his friend Dr. William Darlington, a 19th-Century botanist, physician and politician of Birmingham, Pennsylvania. *Darlingtonia* is a monotypic genus consisting solely of *D. californica*, the only pitcher plant that occurs naturally along and near the west coast of the United States (Figure 46). It is not known whether other species of *Darlingtonia* have existed in the past, but given the diversity of the other genera of Sarraceniaceae, the possibility that greater diversity did exist seems likely. Certainly the formation of the Cascade and Rocky Mountain ranges and the subsequent drying of the interior of the continent could have both eliminated populations of *Darlingtonia* that previously existed in the dessicating interior and restricted the subsequent eastward migration of surviving populations. These circumstances could have inhibited diversification within the genus and possibly contributed to the extinction of species other than *D. californica* if they did indeed exist.

Species Description

Darlingtonia californica Torrey

Original description: Torrey, J., 1853, *Smithsonian Contributions to Knowledge* 6(4): 5.

The specific epithet *californica* refers to the state of California where this plant was originally discovered, although it grows naturally in Oregon too. *D. californica* was discovered in 1841 to the northwest of Mount Shasta by an American botanist, J. D. Brackenridge. Brackenridge was

Figure 46 (facing page). The spectacular foliage of a population of *Darlingtonia californica* growing in dense vegetation in western Oregon.

at the time undertaking an expedition to the remote western frontier and found *D. californica* by chance. The extraordinary shape of the leaves of *D. californica* are widely said to resemble striking cobras poised upright, with red fangs showing, ready to attack. This appearance has given rise to the plant's appropriate common name, the cobra lily.

Plant Structure

D. californica is a clump-forming herbaceous perennial that grows terrestrially along a branching woody rhizome. The foliage is assembled in a compact circular rosette and consists of 3–14 individual leaves. The roots are brown, fibrous and 10–25 cm long. Mature plants readily produce 20–80-cm-long stolon runners which spread underground and develop into separate plants, a trait that enables rapid colonization of surrounding habitat.

The flower of *D. californica* is borne individually on a 25–90-cm-tall scape. Each mature plant typically produces a single flower each spring. The flower consists of 5 yellowish-green, lanceolate, 3–5-cm-long, glabrous sepals and 5 dark-red-to-purple, laterally veined, 2–3-cm-long, lanceolate-oblong petals (Figure 47). The centre of the flower is

Figure 47. An artist's rendering of the flower of *Darlingtonia californica.*

Figure 48. The mature upturned seed pod of *Darlingtonia californica.*

dominated by a large, 5-chambered, 1–2-cm-long, green ovary. The star shaped stigma, which has 5 "arms," extends from the end of the ovary. Some 15–25 stamens are arranged around the base of the ovary in a single series, the anthers are 2–3 mm long, oblong and yellow in colour. The scape is 10–25 cm taller than the mature leaves. The flower of *D. californica* opens between April and June, depending on location; flowering is delayed by several weeks in the more elevated parts of the range of *D. californica*. One to two days after the flower opens, the stigmas become receptive and the anthers shed the pollen, and 6–10 days later the petals and stamen die and fall to the ground. If pollinated, the ovary starts to swell and over the course of 8–14 weeks, it elongates to 3–4 cm in length and develops 40–300 1–3-mm-long, light brown, club-shaped seeds. As the seed pod ripens, it often points upwards and splits open, exposing the seed for dispersal (Figure 48). The surface of the seed is lined with irregular spicules which enable the seed to float and be transported efficiently by the wind.

D. californica produces two types of leaves (figures 49 and 50). During two to three years following germination, young *D. californica* plants produce juvenile leaves which are simple and tubular in shape. The end of the leaf is cut away to form a broad opening underneath an elongated hood structure reminiscent of *S. minor*. The juvenile leaves are generally 1–3 cm in length and are typically reddish green or pure red in colour. In the second or third year of growth, *D. californica* spontaneously produce adult pitchers, the mature leaves of which are the most complex traps of all the pitcher plants. Each leaf is 20–80 cm in height and consists of an upright hollow shaft that terminates in a roughly oval, bulbous, dome structure that extends forward (Figure 51). On the underside of the forward projecting portion of the dome is a 10–20-mm-wide hole located immediately behind a long, pure red, fishtail-shaped appendage. The surface of the dome and the upper parts of the tubular shaft are lined with small, irregular, transparent fenestration. The small hole on the underside of the dome represents the only opening into the interior of the hollow leaf (Figure 52). The perimeter of the entrance hole protrudes into the dome and roles back on itself, forming an inward-projecting rim similar to the entrance of a lobster pot. Nectar is secreted profusely on the back side of the fishtail appendage and around the entrance hole. The interior of the dome is glabrous, smooth and relatively waxy. Downwards-pointing hairs 2–8 mm in length line the interior surface of the pitcher from about one-third of the height of the pitcher to the base of the leaf. The hairs lengthen towards the bottom of the pitcher.

Trapping Process

A strong sweet scent emanates from the profuse secretions of nectar concentrated on the backside of the fishtail appendage of the *Darlingtonia* leaf. The sugary aroma attracts various insects, especially wasps and flies, which climb towards the entrance hole to reach the densest nectar secretions. By following the nectar trails, insects are manoeuvred directly below the entrance hole. From this position, the insect can detect light shining through the transparent fenestration which the insect interprets as the direction of clear sky and freedom (figures 52 and 53). It climbs or flies through the inward-pointing entrance hole, collides with the transparent fenestration, falls down the back of the dome into the vertical tubular shaft and plummets into the liquid contained within. Unable to scale the slippery, waxy interior surface of the leaf and hindered by the downwards-pointing hairs, the insect eventually drowns, is decomposed and the nutrients released from its remains are absorbed by the plant. Mainly flies, beetles and ants are caught.

It remains unclear whether *D. californica* secretes enzymes directly or whether it relies on secondary organisms to break down the insect prey trapped within the pitchers. The acidic liquid within the leaves supports a multitude of bacteria, protozoa and various insect larvae. The interaction between *D. californica* and these microinhabitants is complex, and intricate food webs have been observed and documented. Each species of microorganism performs a different role in breaking down the trapped prey. Larger larvae, such as *Metriocnemus edwardsi,* the larvae of a midge, forage and consume trapped dead insects; the leftovers and their waste are consumed by smaller organisms such as *Sarraceniopus darlingtoniae,* a slime mite; and lastly, microscopic organisms breakdown whatever remains. At each stage, simple compounds are released as waste products which the plant efficiently absorbs. One of the problems posed in understanding the processes of digestion in *D. californica* and all other species of pitcher plants results from the complexity of isolating enzymes in the liquid contained within the leaf pitcher. It remains difficult to establish whether isolated enzymes are produced by the plant or originate from the trapped prey or the organisms that live within the leaf wells.

Distribution

D. californica is distributed across the west coast of Oregon and northwest coast of California. Scattered populations occur inland in mountainous central and southern Oregon and northern California eastward to the Sierra Nevada (Figure 54). *D. californica* occurs from sea level to at least 2400 m above sea level. The growth of upland plants lags about 6–8 weeks behind lowland populations due to the prolonged cold temperatures at higher elevations in spring.

Habitats

D. californica is a highly versatile species. It displays a clear affinity with cool, humid, wet, well-lighted environments and has adapted to inhabit a wide range of wet, boggy, niche habitats throughout the highlands and lowlands of Oregon and California. While the habitat of *D. californica* is traditionally associated with running fresh water and

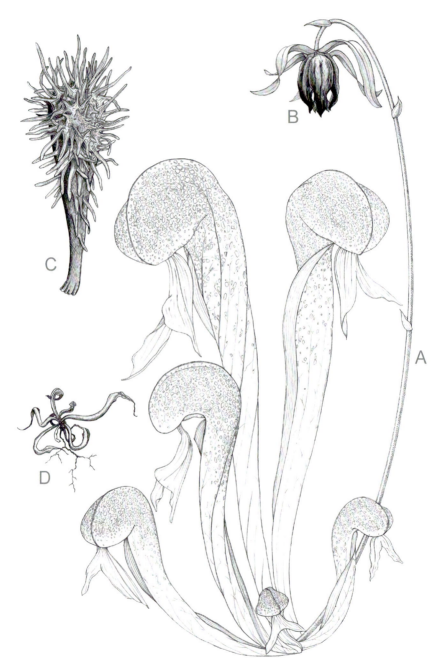

Figure 49. A simplified *Darlingtonia californica* plant, with (A) scape, (B) flower, (C) seed and (D) seedling also shown.

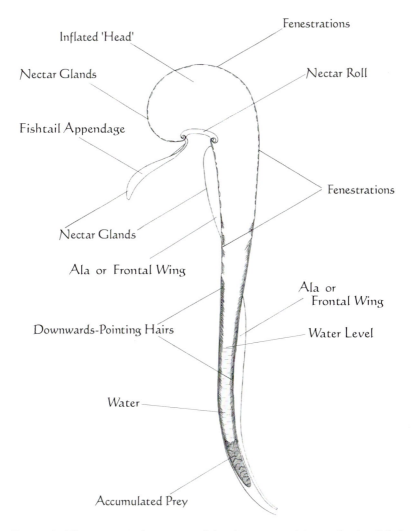

Inflated 'Head'

Fenestrations

Nectar Glands

Nectar Roll

Fishtail Appendage

Fenestrations

Nectar Glands

Ala or Frontal Wing

Ala or
Frontal Wing

Downwards-Pointing Hairs

Water Level

Water

Accumulated Prey

Figure 50. The anatomical structure of the above-ground parts of a simplified *Darlingtonia californica* plant, with pitcher contents added.

Figure 51 (following pages). The mature leaf of *Darlingtonia californica* in (left) front profile and (right) lateral profile.

Figure 52. (Left) The entrance hole to the trap of *Darlingtonia californica* is inconspicuously located on the underside of the forward-projecting dome-shaped portion of the leaf. (Right) Note the transparent fenestration present on the upper parts of the leaf, seen here as viewed from below and within. This feature plays a deceptive role in trapping prey.

Figure 53. The transparent fenestration on the dome of the leaf as seen from above and outside the dome.

Pacific
Ocean

Washington

Oregon

California

Range of
Darlingtonia californica

Figure 54. The range of *Darlingtonia californica* along the northwest coast of the conterminous United States.

serpentine soils, the cobra lily actually inhabits a range of very different habitats.

Along the lowlying coastal areas of Oregon and northern California, populations of *D. californica* grow in open clearings within the mossy, humid, temperate coniferous rainforests (figures 55 and 56). The high precipitation (up to 2300 mm annually) and moderated temperature range affords a lush undergrowth and ground cover of ferns and shrubs within the mossy forests of tall spruce, red cedar and Douglas fir. Permanent populations of *D. californica* generally occur only at open stream sides or in permanently wet sphagnum bogs or marshy spring-fed depressions where the abundance of water inhibits competing vegetation, including trees, and enables more open, sunny habitat. In locations where 0–30% shade and the wet substrate coincide, *D. californica* grows most vigorously and readily forms solid mat-like colonies of plants several meters in diameter. These large, dense clumps are formed by the ability of *D. californica* to reproduce through stolons and maintain dominance over competing bog vegetation. In slightly shaded habitat where taller bog plants grow, the surrounding vegetation plays a significant role in restricting the population size of *D. californica*. *D. californica* has evolved a tolerance of shade and is able to grow amidst light undergrowth in up to 40%-shaded

Figure 55 (following pages). Open, sunny clearings within the mossy temperate rainforest of California and Oregon are home to many populations of *Darlingtonia californica*.

Figure 56. The mossy temperate rainforest of California and Oregon are home to many populations of *Darlingtonia californica*.

conditions, providing adequate water is available. *D. californica* populations in shaded environments are more sparsely distributed and the growth of each individual is less vigorous. In 40–50% shade, the leaves of *D. californica* are etiolated, growth is weak and populations eventually die out.

During the growing season, which occurs from late spring to early autumn, the temperate rainforest habitat of *Darlingtonia* is climatically very similar to the humid, mossy cloud forest that cloaks the lower levels of the tepuis, the habitat of various species of *Heliamphora* in the Guiana Highlands. During the summer, the daytime temperature rarely exceeds 28°C and, more usually, the daily maximum is around 24°C. Humidity is high and precipitation is frequent. Winters are naturally much colder, but the clearings in the coastal rainforests are the mildest of all the *D. californica* habitats. In some localities *Darlingtonia* populations are located just a few hundred meters from the Pacific Ocean, and this, combined with the insulating effect of the tall rain forests, moderates the severity of winter and consequently the coastal forests rarely receive heavy snow or temperatures below -10°C for prolonged periods of time. The substrate of this habitat consists of a mix of acidic coniferous leaf litter and peaty loam. The availability of water varies greatly with location and season; while some populations of *D. californica* grow in permanent standing water throughout the year, those in other locations, particularly upland areas, grow in relatively well drained but permanently moist substrates. Where insufficient moisture is present, the size of *D. californica* plants is stunted and the rate of growth is slowed considerably.

At higher elevations within its range, *D. californica* typically grows at springs and seepage sites overlying serpentine soils (Figure 57). The vegetation at the higher altitudes is generally sparse and *D. californica*

Figure 57. An upland population of *Darlingtonia californica* growing at a wet seepage point.

frequently grows in open, boggy clearings amidst low growing grasses with little or no shade. The temperature range of this habitat is greater than in the lowlands; the winters are colder, they last longer and subzero temperatures, snow and ice typically persist for several weeks. During summer, despite the higher altitude, the lower humidity and reduced vegetation cover result in higher average ground temperatures, but water often slowly trickles or seeps through the substrate and over the roots of *D. californica* and keeps the rhizomes of the plants cool.

Throughout its range, *D. californica* frequently grows along the banks of rushing streams and small lakes amidst tall marsh grasses (figures 3 and 58). It competes effectively with the surrounding vegetation and often only the tops of the pitchers are visible amidst the undergrowth. In winter, riverside habitats are frequently flooded and often remains inundated for several weeks.

General Ecology

The colouration of *D. californica* responds directly to the ambient light levels of its habitat. In 5–20% shade, the leaves are yellowish green, the fishtail appendage is pure red or purple and, in some strains, a variable degree of light red venation is present on the lower parts of the dome. In 25% or more shade, the lower light intensity stimulates the plant to increase the level of chlorophyll within its leaves to compensate for inadequate photosynthesis (Figure 59). As a result, the non-photosynthetic red pigmentation in the leaves is minimized and the size of the translucent fenestration is reduced, causing the foliage to appear a darker and more uniform apple green in colour. In very bright conditions, *D. californica* responds by reducing the chlorophyll content of the leaves and increasing the fenestration size which causes the pitchers, especially the dome-shaped tops, to appear yellow in colour. Red colouration and veining on the dome and fishtail appendage are also intensified in bright conditions (Figure 59). These differences in colouration are environmentally induced and would change if the local growing conditions were altered.

A degree of genetic variation is discernable in *D. californica*. Some strains consistently develop intense reddish-purple colouration that

Figure 58 (facing page). A population of *Darlingtonia californica* growing on the bank of a small stream.

Figure 59. The effect of different lighting conditions on *Darlingtonia californica:* (left) an etiolated individual growing in 40% shade and (right) an intensely coloured individual growing in direct sunlight.

dominates all parts of the upper half of the leaves (figures 60 and 61), but it is difficult to fully appreciate the depth of this variation since this red colouration is masked in populations that occur in shaded habitat. The leaves of juvenile *D. californica* typically suffuse pure red in colour, especially during winter, but this colouration is lost as the plant matures.

One distinctly coloured natural variety of *D. californica* has been identified. A strain of *D. californica* which altogether lacks the ability to produce red pigmentation was discovered by Dr. Barry Rice and named as the cultivar 'Othello.' The leaves and flowers of this variant are consistently pure yellowish green and do not display any red colouration regardless of the conditions in which it grows (Meyers-Rice 1997, 1998).

The growth of *D. californica*, like that of *Sarracenia*, is defined by the seasons and follows an annual cycle. During March, or April in the upland parts of the species' range, *D. californica* emerges from winter dormancy, flowers from April until June and then produces carnivorous pitchers at a constant rate throughout the remainder of the growing season

Figure 60. A strain of *Darlingtonia californica* exhibiting prominent red colouration.

(figures 46 and 55). During October, prompted by the colder temperatures and shorter days, *D. californica* stops growing, becomes dormant and stays so through the remaining autumn and the winter. Foliage on the dormant plant remains green until the following spring, then slowly dies back over the course of the next growing season.

Nectar is secreted profusely on the pitchers of *D. californica* which creates a distinctly sweet scent that is discernable from a distance of several meters, especially in areas where large clumps of plants grow together. The aroma is similar to that of *Heliamphora* and is less sweet than the scent of *Sarracenia*. Like the lids of *Heliamphora* and *Sarracenia*, the fishtail appendage of *D. californica* is adapted to distribute the aroma of nectar and serves the purpose of attracting insect prey. As the

Figure 61. This strain of *Darlingtonia californica* exhibits unusual red venation on the upper parts of the tubular portion of its leaves.

bait in the trap, the nectar secretions are concentrated on the fishtail appendage and advertised accordingly by intense red colouration. *Heliamphora* and *D. californica* are better adapted to growing in shaded conditions than *Sarracenia* and the two genera rely to a greater degree than Sarracenia on asexual reproduction, especially in overgrown habitat where dense vegetation prevents seed from germinating. The tendency of mature *D. californica* plants to continually divide as they grow is very similar to the same tendency seen in *Heliamphora*, especially *H. chimantensis*. Old populations of *D. californica* form solid mats of plants several meters across, which in some cases consist of a single genetic individual that has perpetually divided and multiplied. This trait differs from the ability of *Sarracenia* to form small clusters of plants that result from offshoots. Since 3–7 years are required for *D. californica* to grow from seed to mature plants, asexual reproduction in this species represents a faster and more assured method of reproduction.

Many aspects of the leaves of *D. californica* are altogether unique among carnivorous plants. The transparent fenestration differs completely from the opaque white areolation of certain *Sarracenia* species even though the underlying function of the two characteristics is essentially the same. As the leaves of *D. californica* grow and inflate, the upper parts, including the dome-shaped top, twist counterclockwise by 90–270° so that the 5–20-mm-wide ala uniquely winds around the tubular shaft of the leaf. Furthermore, *D. californica* is the only pitcher plant which has evolved the ability to reproduce by stolons (Figure 62). The runners, readily produced as soon as a healthy plant reaches maturity, extend underground and spread away from the mother plant for up to 100 cm, then eventually grow into a separate individual plant. This ability enables *D. californica* to rapidly colonize surrounding habitat and maintain dominance against competing vegetation.

Figure 62. A stolon spreading from a *Darlingtonia californica* plant. Here, the stolon is growing above ground due to the thin, stony substrate.

Heliamphora

The name *Heliamphora* is derived from the Greek words *helos* (marsh or meadow) and *amphora* (pitcher or urn), thus the *Heliamphora* are the marsh pitcher plants (Figure 63). Fifteen species have been described to date and several more taxa await classification. *Heliamphora* are known primarily from the cool, marshy summits of the great sandstone mesas of southern Venezuela, Guyana and the extreme north of Brazil. These spectacular table mountains, locally known as tepuis or tepuyes, rise hundreds of meters above the tropical, hot rainforests and dry savannahs of the lowlands and are lost in the clouds above. This region, known as the Guiana Highlands (Figure 18), is among the most inaccessible and least explored corners of the planet and, consequently, the majority of *Heliamphora* species have been named only during the past few years.

Plant Structure

The *Heliamphora* are clump-forming, perennial plants that produce hollow infundibular or cylindrical leaves. The foliage is assembled in a circular rosette and consists of 3–10 individual leaves. Most *Heliamphora* species grow along a decumbent branching rhizome. The roots are brown, fibrous and 10–25 cm long. Some 1–10 bell-shaped flowers are borne on 20–50-cm-tall, erect scapes. The flower of *Heliamphora* (Figure 64), very similar in all species other than *H. neblinae*, consists of four lanceolate, 3–6-cm-long tepals that are white, pink or red, depending on their age. Some 8–24 stamens are present in 1 or 2 series (Figure 65). The number, shape and size of the stamens differ among species and, in some cases, possibly also among local populations

Figure 63 (facing page). The purple black variant of *Heliamphora pulchella* has spectacular colouration.

Figure 64. The typical flowers of *Heliamphora*.

within a species. The anthers are oblong, 3–11 mm in length, 1–3 mm wide and yellow in colour. The peduncle and the pedicels are glabrous in all species except *H. neblinae*. The pedicel is 2–5 cm long. The bracts are ovate or lanceolate and 3–5 cm in length. The ovary is 3 celled. The seed is circular to ovate in shape and golden yellow in colour. Each seed has an irregular 1–3-mm-wide wing which enables it to be distributed effectively by the wind. The seed also floats and is likely further dispersed during occasional flooding of habitat.

The different parts of the *Heliamphora* flower ripen at different times to prevent self-pollination. When the flower first opens, the stigma is unreceptive, the anthers are not ripe and the tepals are pure white in colour. Within three days after opening, the stigma ripens and becomes receptive to pollen which is brought to the flower by insect pollinators, particularly beetles and wasps. The anthers remain unripe and dull yellow or green in colour. Then, 2-3 days later, the end of the stigma turns brown and becomes unreceptive to pollen. Only when the stigma is unreceptive do the anthers finally ripen, turn bright yellow and have pollen available for visitors.

The design of the *Heliamphora* flower is extremely resourceful (Figure 66). When ripe, the anthers do not split open and freely discharge

Figure 65. The reproductive organs of a *Heliamphora* flower.

pollen as occurs in flowers of most plants. Instead, only a small hole at the end of each anther opens. The pollen can only be obtained by specific pollinators that regularly visit the flowers of *Heliamphora* and reliably deliver the valuable pollen to the next *Heliamphora* plant. It is the low-frequency buzzing vibration of the beating wings of wasps and certain beetles that is the key to unlocking the pollen, and the vibrations which these insects create cause the pollen to spurt out of the anthers and stick to the insects' bodies, ready to be delivered to the stigma of another flower. In cultivated plants, the pollen can be released from the anthers by the vibrations of a tuning fork and transferred to other flowers with a small brush.

At 4-5 days after the flowers open, the stamens die and fall away and the tepals slowly suffuse pinkish red in colour. If fertilized, the ovary swells and, over the course of the following 6–12 weeks, 10–100 seeds develop. As the flower ages, the tepals continue to darken and eventually reach a shade of dark red when the seeds are ripe. Often, but not always, the seed pods bend upwards so that the seed is effectively blown away in the wind (Figure 67). Despite the efficiency by which *Heliamphora* seed is transported, most species are highly localized and occur on just one or two mountains. All *Heliamphora* species are inter-fertile and self-fertile. Different species readily hybridize where territories naturally overlap

Figure 66. The flowers of *Heliamphora elongata*.

and, occasionally, complex crossbreeds also occur in the wild. *Heliamphora* species produce seed most prevalently when different strains, or different species, cross-pollinate.

Often the first bract on the flower scape differs from the regular ovate or lanceolate shape and instead takes the form of a small, rudimentary pitcher. This characteristic occurs apparently randomly on the flowers of all *Heliamphora* species and does not appear to serve any specific purpose. The bract-pitchers of flowers in all species of *Heliampora* are 5–15 cm long and take the form of mature leaves of each species (Figure 68). Only one bract pitcher is ever present on the flowers of *Heliamphora*. The bract pitchers rarely catch prey.

Heliamphora produce two types of leaves. During the first 2–4 years after germination, young plants produce juvenile leaves which are simple

Figure 67. The up-turned seed capsules of *Heliamphora minor.*

and tubular in shape and typically up to 12 cm in length (Figure 69). The end of the juvenile leaf rolls back on itself to form a small, narrow opening and the upper part of the leaf curves forward slightly. The juvenile leaves vary only slightly among the various *Heliamphora* species. Typically the juvenile leaves are red, pink or orange in colour and are generally too narrow to catch any prey. During the third to fifth year of growth, the juvenile plant spontaneously converts to producing adult pitchers.

The mature leaves of *Heliamphora* species are infundibular or cylindrical in shape and are considerably larger and more voluminous than the juvenile leaves. In all species with the exception of *H. sarracenioides*, a conical or concave structure, known as the nectar spoon, is present at the apex of the leaf. The interior side of this structure is lined with nectar-secreting glands which slowly exude droplets of the sweet liquid to

Figure 68. The bract-pitcher on the scape of *Heliamphora chimantensis.*

attract insects. In some species, nectar is secreted so profusely that it drips down the interior of the leaf and creates a very sweet aroma which is discernable from several meters away. The exterior of the upper and middle sections of the leaves are sparsely lined with 1–2-mm-long, red nectar glands (Figure 70) which attract ground-dwelling insects, especially ants, and encourage them to climb up the outside of the leaf to the entrance of the trap.

In many species, the interior of the upper section of the leaf is lined with spine-like downwards-pointing hairs, although this characteristic varies greatly among the respective *Heliamphora* species and is altogether lacking in several species. In all species of *Heliamphora,* the level of water contained within the leaves is regulated, either by a small drainage slit (in *H. chimantensis, H. minor* and *H.*

Figure 69. The juvenile leaves of *Heliamphora.*

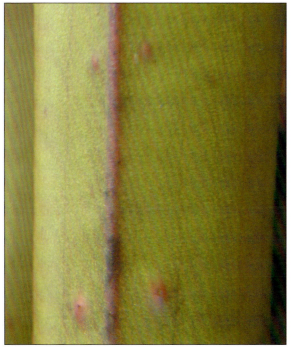

Figure 70. Nectar glands on the exterior of the leaves of *Heliamphora*.

pulchella) or a narrow drainage hole present halfway up the front side of the leaf (in all other species) (Figure 71). These features control the water level within the leaves and prevent the pitchers from filling up with rain water and toppling over. A patch of interlocking, long, downwards-pointing hairs line the interior side of the leaf around the drainage hole and prevent small insects, once trapped, from climbing through it and escaping.

In most species, just above the water line, the girth of the leaf is restricted and forms an inward overhang that makes it is difficult for trapped insects to escape (figures 72 and 73). The interior surface of the leaves is extremely waxy and, in all species other than *H. folliculata*, a ring of 5–12-mm-long, downwards-pointing hairs is present just above the water level which prevents insects from scaling the interior of the leaf and escaping. The interior surface of the lower section of the leaves of all *Heliamphora* species is glabrous, smooth and lined with specialized glands which absorb nutrients released from trapped prey. The interior surface of the very bottom of the leaf is sparsely lined with 2–5-mm-long

Figure 71. The (left) drainage hole and (right) drainage slit in *Heliamphora*.

hairs. The purpose of these hairs is not clear as they are located several centimeters below the water line.

Trapping Process

The vibrant colouration and profuse secretions of nectar attracts various insects to the leaves of *Heliamphora*. Bees, wasps, flies and ants are the most common visitors, but butterflies, beetles and craneflies also visit and are regularly caught. Once attracted to the pitcher opening, the visiting insects detect the heaviest concentrations of nectar oozing from the nectar spoon on the back of the leaf and scale the slippery interior of the trap to reach it (Figure 74). The inner surface is waxy and, in most species, lined with downwards-pointing hairs which ensure that a foot-hold is difficult to establish and maintain (Figure 75).

As the insect manoeuvres itself to reach the nectar spoon, it clings to the vertical interior surface of the leaf in an ever more precarious way. The slightest falter causes the insect to plummet straight down into the depths of the trap and the digestive liquid contained within. Once inside, the unfortunate victim is unlikely to escape. Winged insects in particular soon become waterlogged and hapless. A further set of long, spine-like, downwards-pointing hairs pose a formidable obstacle for insects trying to climb the slippery interior of the leaf to safety; trapped and unable to escape, the prey dies of exhaustion or drowns. The remains of trapped insects accumulate within the leaves (Figure 76). Even though competition for attracting insects is high, each pitcher typically and regularly catches several victims.

The role of enzymes and secondary organisms in the digestive process of *Heliamphora* is not fully understood. Mosquito larvae and various other micro-organisms readily inhabit the traps of all *Heliamphora* and actively assist the digestion process by scavenging the remains of the trapped victims. In doing so, the remains of prey are broken down and simple nutrients are released into the digestive liquid within the leaves and are thereby available to be absorbed by the plant. It remains unclear whether digestive enzymes are or are not secreted directly by the plant.

Distribution

Heliamphora are distributed across the south of Venezuela and the borderlands of northern Brazil and western Guyana (Figure 77). The topography of this region is dominated by colossal table mountains, tepuis, which rise abruptly hundreds of meters above the surrounding lowlands. The majority of *Heliamphora* species occur, exclusively or predominantly, on the summits of the tepuis which lie at 1800–2800 m above sea level, and many species are endemic to only one or two of these summits. Several species of *Heliamphora* occur amidst the cloud forest of the foothills and flanks of the tepuis, and a minority of these species occur in wet marshy clearings of the tropical lowlands. The majority of all *Heliamphora* species occur within the Gran Sabana, a region in the south of Venezuela at the heart of the Guiana Highlands, although several species also occur on much more isolated, outlying mountains across Bolivar State in southwestern Venezuela.

Habitats

Heliamphora typically occur in three general habitats — the plateau summits of the tepuis, the cloud forests that occupy the flanks and foothills near the base of the tepuis and the wet areas of lowlands that lie beyond the foothills of the tepuis.

Figure 72. A simplified *Heliamphora* plant, with (A) mature leaf, (B) scape, (C) seed pod, (D) seeds and (E) seedling also shown.

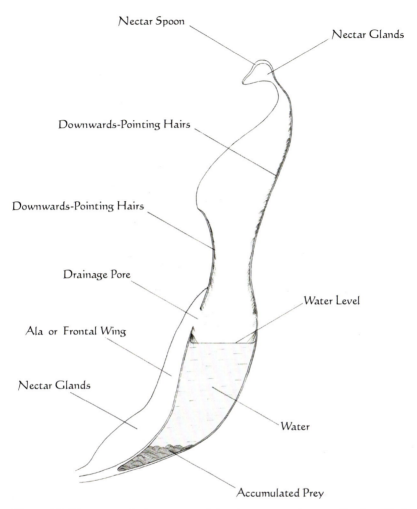

Nectar Spoon

Nectar Glands

Downwards-Pointing Hairs

Downwards-Pointing Hairs

Drainage Pore

Water Level

Ala or Frontal Wing

Nectar Glands

Water

Accumulated Prey

Figure 73. The anatomical structure of the above-ground parts of a simplified *Heliamphora* plant, with pitcher contents added.

The plateau tops of the tepuis are among the wettest places on Earth and receive intense rainfall that totals up to 9000 mm annually. The abundance of water is the restricting factor for life in this habitat and, consequently, the tepui summit environment is classified as a rain desert (Figure 78). The incessant rains scour the mountain surfaces of sediment and wash away organic matter and nutrients before soil can accumulate. Up to 95% of the surface of the tepuis is bare rock and in this barren, inhospitable landscape, vegetation is scarce and generally

Figure 74. A wasp eating nectar from a variant of *Heliamphora pulchella* that lacks the long downwards-pointing hairs found in others of the species.

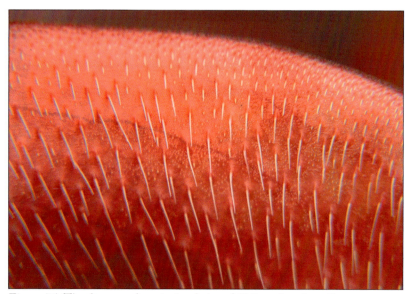

Figure 75. The downwards-pointing hairs of *Heliamphora ionasii*.

Figure 76. The contents of a pitcher of a *Heliamphora nutans* plant.

restricted to the gullies and channels in the rock surface where a little acidic (pH 3–5), peaty or sandy substrate collects and affords a rare roothold. It is in this harsh setting that *Heliamphora* flourish, partly through their unusual ability to procure nutrients through carnivory.

The summits of the tepui mountains are cool, ultra-humid, highland environments (figures 79 and 80). The daytime temperature on most of these surfaces typically peaks between 16–23°C and drops to 7–12°C during the night, a range that remains relatively constant throughout the year. Only on the highest tepuis (Mount Roraima, Kukenán Tepui and Mount Neblina) does the nighttime temperature fall significantly lower and occasionally reach an absolute minimum of 2–3°C. Even in these extreme locations, however, subzero temperatures have never been recorded. Temperatures in excess of 26°C are unusual and generally restricted to low-lying plateaus that are influenced by the lowland climate. Annual precipitation varies significantly across the Guiana Highlands, but all of the tepuis receive abundant rainfall of between 6000 mm and 12,000 mm annually, generated mainly through orographic processes. The tepuis of Bolivar State, Venezuela, are generally taller, wetter, more

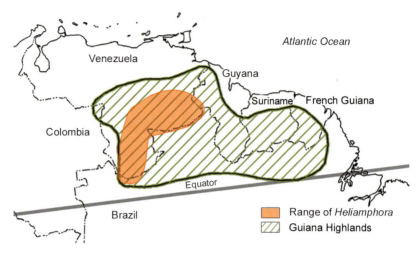

Figure 77. The range of *Heliamphora*.

desolate and support substantially less vegetation than those of Amazonas State, Venezuela. All of the habitats of *Heliamphora* are located within 8° north of the Equator and are exposed to intense tropical sunlight. The granular substrate on the summits of the tepuis is mostly thin, saturated, peaty and acidic. *Heliamphora* most frequently root and grow in organic substrates, but occasionally they grow in those that consist predominantly of sand and coarse gravel.

Many *Heliamphora* also occur in open areas of the cloud forests that cover the lower flanks and foothills several hundred meters below the tepui summits. At least one species, *H. ionasii*, occurs exclusively within this habitat (Figure 81). The environment at this lower altitude, generally 1200–1800 m above sea level, is more hospitable and several degrees warmer than that of the tepui summits. The daily maximum temperature peaks between 22–27°C and the nightly minimum ranges between 8–16°C. Water is even more abundant within the cloud forest than on the tepui summits. Warm, humid air masses that rise against the tepuis shed often copious amounts of rainfall as they ascend, all of which combines with the runoff from the towering tepui summits above to provide abundant moisture to the slopes surrounding the bases of the tepuis. Particularly after heavy thunderstorms on the plateau tops, chains of waterfalls flood over the cliff sides and saturate the hills below (Figure 19). Furthermore, during the cool nights, a heavy dew descends upon the landscape and saturates the area by morning.

Figure 78. The barren, inhospitable habitat of *Heliamphora minor.*

These conditions support dense cloud forests which carpet the foothills of the plateaus. Most of the foothill-dwelling *Heliamphora* occur only in open marshy clearings or in open-canopy cloud forest on ridge sides or marshy plains below the waterfalls. The substrate in these areas consists entirely of waterlogged humus that directly overlies the bedrock. The marshy soil is considerably more nutrient-rich that the granular sediments of the tepui summits.

Away from the clearings, the cloud forest is extremely dense and impenetrable, but a number of *Heliamphora* species are adapted to surviving within this forest and grow in open areas on the forest floor in 15–60% shade. *H. ionasii* grows most frequently in this environment and competes efficiently for light with the surrounding vegetation. In these shady conditions, it produces massive, non-etiolated pitchers as large as 45 cm in height.

Figure 79 (following pages). The magnificent Mount Roraima is an immense sandstone plateau that lies directly upon the point where the borders of Brazil, Guyana and Venezuela meet.

A few species of *Heliamphora* also occur in the lowlands surrounding the tepuis at altitudes of 600–1000 m above sea level (Figure 82). In most cases, lowland populations occur at poorly drained seepage areas where the substrate remains permanently moist or saturated. This habitat is much hotter and considerably less humid than the plateau summits and side hills. Here, the daily maximum temperature peaks between 24–28°C and regularly exceeds 30°C during the dry season. The nightly minimum ranges between 12–18°C. Rainfall is considerably less

Figure 81. *Heliamphora ionasii* growing amidst the dense vegetation of a cloud forest in the Guiana Highlands.

Figure 80 (preceding pages). This awe-inspiring landscape of desolation on the summit of Mount Roraima displays the barren rocky surface containing numerous light-coloured puddles and ponds which are recharged daily by showers and storms.

Figure 82. The low-lying marshy habitat of lowland *Heliamphora* populations.

than at higher elevations, and can be as little as 600 mm in some areas. The lowland populations of *Heliamphora* always occur in the company of *Stegolepis* and *Brocchinia*, often including *B. hechtioides* or *B. reducta*. Generally, lowland *Heliamphora* grow in 0–20% shade and are absent where vegetation is 90 cm or more high. The substrate in these lowland sites consists of a mix of acidic peaty humus and sand. Currently, only *H. heterodoxa* and *H. nutans* are known to occur in the lowlands surrounding the tepuis, but several other species likely also occur in this habitat, in particular *H. minor* and *H. pulchella*, given their respective ranges. The lowland populations of *H. nutans* were first observed by Richard Schomburgk during his 1938–1939 expedition to the base of Mount Roraima, but these have not been revisited for many decades.

General Ecology

Heliamphora are highly adaptable plants and actively respond to the environmental conditions of their habitat in many dramatic ways. In particular, light directly controls many aspects of the colouration and shape of the leaves. In direct sunlight the pitchers of *H. pulchella* are

Figure 83. A variant of the typical variant of *Heliamphora pulchella* growing in full sunlight. Note the long, downwards-pointing hairs on the interior of the plants' leaves.

8–10 cm tall and cylindrical in shape. The interior of the leaf is lined with a variable coating of long, downwards-pointing hairs and the pitchers are typically pure red or dark purple in colour (Figure 83). Each plant consists of a rosette of 4–8 individual leaves.

When *H. pulchella* grows in partial shade, the leaves respond to the reduced light levels and the lower efficiency of photosynthesis by enlarging in size, reaching 12–20 cm in height, and growing more infundibular in shape in order to catch more light. The red and purple pigments of the foliage are replaced with extra chlorophyll and, consequently, the leaves appear predominantly green (Figure 84). The lining of hairs on the interior of the leaves is less dense and the lower input of energy only enables the plant to sustain 3–4 green leaves at any one time. Growth is slower and weaker than would occur in plants that receive more light, and fewer flowers are produced.

In yet darker conditions, the leaves of *H. pulchella* etiolate and the tubular aspect of the trap is altogether lost. The chronic lack of light causes the plant to forsake carnivory and to photosynthesise enough energy to survive. The leaf, as in all etiolated *Heliamphora*, is flat, glabrous and pure green in colour (Figure 85). The nectar spoon is not formed and the leaf bears little resemblance to the foliage of *Heliamphora* that grow in bright conditions.

All *Heliamphora* respond in this way to light intensity. It is therefore extremely difficult to identify the species represented by specimens growing in partial shade since the characteristics of each species are masked through the etiolation of the leaves. However these responses are symptoms of environmental conditions, and will alter if the conditions of the habitat change. If an etiolated, green *H. pulchella* growing in the dense shade were transplanted to a bright, sunny location, it would adapt to the bright conditions and revert to producing colourful, cylindrical leaves and eventually would appear identical to specimens that had always grown in sunny habitats.

Figure 84. *Heliamphora pulchella* growing in partial (30%) shade.

Figure 85. *Heliamphora pulchella* growing in strong (70%) shade. The leaf is significantly etiolated.

Heliamphora also respond dramatically to drought stress. If little water is available and the substrate gradually becomes increasingly drier, the size of the leaves is reduced in order to minimize the surface area of the foliage through which water is transpired (Figure 86). Dwarfed specimens of *Heliamphora* produce leaves that are just 3–5 cm tall. All *Heliamphora* species can permanently grow and even flower at this dwarfed size, but like the response to light intensity, the dwarfed specimens revert to typical size if environmental conditions normalize. Many *Heliamphora* regularly grow along the banks of small streams and rivers and also develop short, stunted leaves due to their exposure to rushing water and lack of nutrients (Figure 87). In the wild, *Heliamphora* are highly tolerant of flooded conditions, but *Heliamphora* grown in standing water under cultivation often face death since the water stagnates. In cultivation, *Heliamphora* are best grown in well drained substrate and if watered daily.

Unlike *Darlingtonia* and *Sarracenia*, *Heliamphora* have no growing cycle, they produce leaves and flowers at a steady rate throughout the year and do not undergo a period of dormancy. *Heliamphora* raised from seed will mature in 3–7 years.

Figure 86. *Heliamphora pulchella* dwarfed by the lack of water.

Figure 87. *Heliamphora pulchella* growing along a stream bank. Tannin that has leached from the leaf litter is causing the orange colouration of the water.

Species Descriptions

Heliamphora chimantensis Wistuba, Carow and Harbarth

Original description: Wistuba, A., T. Carow and P. Harbarth, 2002,
Carnivorous Plant Newsletter 31(3): 78–82.

The specific epithet *chimantensis* refers to the Chimanta Massif, a group of tepuis in the western part of the Gran Sabana where the species occurs endemically.

The range of *H. chimantensis* is known to include the summits of Chimanta and Apacara tepuis at the centre of the Chimanta Massif (Figure 88). Individuals of the species occur predominantly in open marshy habitats between 1900–2100 m above sea level. This species readily hybridizes with *H. pulchella.*

Although *H. chimantensis* flowers profusely, it seldom produces seed and appears to reproduce primarily through division. Over many decades, individual plants continually produce offshoots and form gigantic densly packed cushions of pitchers which can consist of several hundred individual plants and can measure up to 6 m in diameter (figures 14, 20 and 89). The tendency of *H. chimantensis* to grow in this way is unique among *Heliamphora* species and perhaps is a response to the climatic extremes of its highland environment — much like the cushion plants of New Zealand. *H. chimantensis* clumps consist of leaves that are pressed densly together and collectively form a solid mat rather than the open cluster of offshoots which arise in most other species of *Heliamphora.* Indeed, this growing habitat appears to be more similar to that of *Darlingtonia* than to the other *Heliamphora* species.

The pitchers of *H. chimantensis* are 25–50 cm long and 4–7 cm wide at the pitcher opening (Figure 90). The lower and middle sections

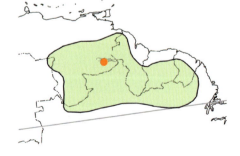

Figure 88. The range of
Heliamphora chimantensis.

Figure 89. A vast clump of *Heliamphora chimantensis* consisting of several thousand individual offshoots.

of the leaves are infundibular. The upper section is cylindrical or slightly infundibular. The girth of the leaf is constricted slightly above the water line. A prominent V-shaped slit extends down the front of the leaf almost to the midsection; this characteristic is more prominent in *H. chimantensis* than in *H. minor* and *H. pulchella*, the other species that

Figure 90. The (left) colourful pitcher and (right) distinctive nectar gland of *Heliamphora chimantensis.*

have drainage slits. No drainage hole is present on the leaves of this species.

The interior of the leaf is glabrous with the exception of a ring of 3–6-mm-long, downwards-pointing hairs present above the water line. Large irregular, dark red nectar glands are present on the inner surface of the nectar spoon (Figure 90). The irregular shape of these glands is unique and helps to distinguish this species. The nectar spoon is oval in shape, slightly concave and pointed towards the apex. Nectar is secreted profusely on the leaves of *H. chimantensis,* and large clumps of this species have a very strong, sweet fragrance that is discernable from several meters away.

The pitchers typically open yellowish green in colour and turn shades of orange and red as they age. Localized populations display different colour traits, some strains being predominantly red in contrast to others that produce largely yellowish-green leaves with red nectar spoons. Faint red veins are present on the upper section of the leaf of some strains. The leaves turn pure maroon or scarlet as they die back.

Heliamphora elongata Nerz

Original description: Nerz, J., 2004, *Carnivorous Plant Newsletter* 33(4): 111–116.

The specific epithet *elongata* refers to the elongated shape of the leaves of this species. Prior to its description as a species, *H. elongata* was referred to as *H. sp. Ilu / Tramen*.

The range of *H. elongata* consists of the summits and foothills of Ilu and Tramen tepuis in the eastern part of the Gran Sabana, 1800–2600 m above sea level, the summit of Yuruani Tepui, 2100–2300 m above sea level; and probably also the small, inaccessible summits of Karaurin and Wadaka tepuis located between Ilu and Yuruani (Figure 91). The lower altitudinal limit of this species within the cloud forest at the bases of Ilu and Tramen tepuis and elsewhere is currently not known. *H. elongata* is known to hybridize only with *H. ionasii* in the valley between Ilu and Tramen tepuis. The populations of *H. elongata* on Yuruani Tepui differ from those on Ilu and Tramen tepuis in that the pitchers of the Yuruani population are smaller and less elongated.

The summits of Ilu, Tramen and Yuruani tepuis are among the most inhospitable and desolate of all the Guianese mesas. Due to the high altitudes of these plateaus, the climate is cold — as low as 3°C during the coldest nights — and especially wet. *H. elongata* has adapted to the hardships of its habitat by growing in the company and mutual shelter of other vegetation, especially orchids, *Stegolepis* and *Bonnetia* (Figure 92). Where it does grow alone, exposed to the elements, the size and vigour of the individual plants is greatly stunted.

The pitchers of *H. elongata* are 10–35 cm tall and 4–7 cm wide at the pitcher opening (Figure 93). The lower and middle sections of the leaf are infundibular and slightly ventricose. The girth of the leaf is constricted

Figure 91. The range of *Heliamphora elongata.*

Figure 92. A colourful clump of *Heliamphora elongata* growing amidst stunted vegetation.

Figure 93. Two small clusters of *Heliamphora elongata* displaying the typical shape and colour of leaves of the species.

slightly above the water line. The upper section of the leaf is narrow, infundibular and elongated. In some strains the angle of the leaf changes at the midsection so the upper parts of the leaf are bent slightly to or away from the centre of the rosette. The interior of the upper section of the leaf is uniformly lined with 1–3-mm-long, downwards-pointing hairs. The hairs are often very shiny. A ring of 2–6-mm-long hairs is present on the interior surface of the leaf just above the water line. The lid is helmet-shaped, but more concave than the lid of *H. nutans*. The pitcher opening dips at the front of the leaf. The leaves typically open yellowish orange in colour and slowly suffuse shades of pink and red as they age (Figure 94).

Figure 94 (following pages). A clump of stunted *Heliamphora elongata* growing amidst a stark and inhospitable landscape of bare rock.

Heliamphora exappendiculata Nerz

Original description: Nerz, J., and A. Wistuba, 2006, *Carnivorous Plant Newsletter* 35(2): 43–51.

The specific epithet *exappendiculata* is derived from the Latin *ex* (out of, or lacking) and *appendicula* (appendage) and refers to the apparent lack of a nectar spoon lid in this species. *H. heterodoxa* (Steyermark) var. *exappendiculata* (Maguire) (1978) and *H. heterodoxa* var. *exappendiculata* fm. *glabella* Steyermark (1984) are taxonomically invalid and synonymous with *H. exappendiculata*. *H. exappendiculata* is known to hybridize only with *H. pulchella* where the two species are found growing together in parts of the western tepuis.

The range of *H. exappendiculata* encompasses the Chimanta Massif, Aprada Tepui and Araopan Tepui in the western sector of Bolivar State, Venezuela (Figure 95). The known altitudinal range of *H. exappendiculata* is 1700–2100 m above sea level, although it is possible that the species also occurs on the inaccessible cliff sides and amidst the cloud forest of the foothills of the western tepuis at considerably lower altitudes.

H. exappendiculata frequently grows rooted directly to the sides of bare, mossy cliffs of narrow valleys and river gorges, especially where waterfalls and seepage ensures that the rock face is permanently wet. *H. exappendiculata* occurs most frequently in 10–40% shaded conditions and appears to prefer poorly lit habitat. Under conditions of 30% shade, the leaves of this species are pure green with the exception of the nectar glands which are red in colour. Under 50% or more shade, the leaves of *H. exappendiculata* are etiolated and altogether green. Occasionally this species grows in open, sunny habitat, and produces leaves that are a beautiful pure shade of orange or light red.

Figure 95. The range of *Heliamphora exappendiculata*.

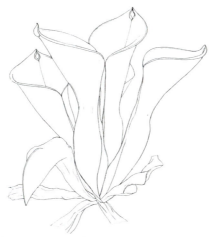

Figure 96. The unique structure of the leaves of *Heliamphora exappendiculata*.

The pitchers of *H. exappendiculata* are 15–25 cm tall and 3–6 cm wide at the pitcher opening (Figure 96). The lower section of the leaf is infundibular, but the girth is restricted at the midsection which causes the lower portion of the leaf to appear slightly ventricose in profile. The upper part of the pitcher is infundibular, especially towards the pitcher opening. Uniquely, the apex of the leaf does not form a "nectar spoon lid" as in all other *Heliamphora* species but, instead, nectar glands are set into the back wall of the pitcher and take the form of small, concave, bubble-like structures. The apex of the leaf is rounded but terminates in a slight point. The interior of the upper section of the leaf is lined with 1–3-mm-long downwards-pointing hairs and, in some strains, the exterior of the leaf is slightly pubescent. Perhaps in compensation for the diminutive apex nectaries, the exterior of the leaf is lined with up to 30 1–2-mm-long, circular, red nectar glands which efficiently attract ground insects and other prey.

Heliamphora folliculata Wistuba, Harbarth and Carow

Original description: Wistuba, A., P. Harbarth and T. Carow, 2001, *Carnivorous Plant Newsletter* 30(4): 120–125.

The specific epithet *folliculata* is derived from the Latin *folliculus* (small bag or purse) and refers to the distinctive bubble-like nectar-storing structure present on the back of the pitcher.

Figure 97. The range of
Heliamphora folliculata.

The range of *H. folliculata* consists of the summits of the remote Los Testigos chain of tepuis — Aparaman Tepui, Murosipan Tepui, Tereke-Yuren Tepui and Kamarkaiwaran Tepui — located in the northern part of the Gran Sabana (Figure 97). The species occurs between 1700–2100 m above sea level. The summits of the Los Testigos tepuis are small and exposed to harsh climatic conditions, and like *H. elongata*, *H. folliculata* becomes established mainly in the shelter of low-growing grasses in well drained, sunny locations on the mountain summits. In some areas it grows directly on rock overhangs on the cliff sides of the tepuis. *H. folliculata* is isolated from all other species of *Heliamphora* and, as a result, no naturally occuring hybrids are known.

The pitchers of *H. folliculata* are infundibular in the lower section and cylindrical in the upper and middle sections (figures 98, 99 and 100). The girth of the midsection is constricted only very slightly so that in cross section the leaf is more tubular than in other *Heliamphora* species. The pitchers are between 15–30 cm tall and roughly 3–6 cm wide and 2–4 cm deep. The leaf is broader than it is deep and the front is always pressed inward slightly so that the cross section of the leaf is reniform. The nectar spoon forms a small, hollow chamber which stores nectar, and it is often torn open by some unknown animal — perhaps a bird or a wasp. The nectar chamber is between 5–10 mm in length and curves forward slightly towards the front of the leaf (Figure 99).

The uppermost 10–20 mm of the interior of the pitcher is lined with fine 1-mm-long, downwards-pointing hairs. The rest of the interior of the leaf is glabrous except for a small patch of 4–5-mm-long,

Figure 98 (facing page). A colourful clump of *Heliamphora folliculata* growing amidst short grasses.

Figure 99. Two views of the nectar chamber of *Heliamphora folliculata* leaves: (left) oblique and (above) vertical from above.

Figure 100. A cluster of *Heliamphora folliculata* showing typical colour and shape of the leaves.

downwards-pointing hairs which surround the interior of the drainage hole and prevent any small trapped prey from escaping. Uniquely, no ring of long downwards-pointing hairs is present at the water line of this species. In some strains, the exterior of the leaf is lined with a sparse coating of fine 0.5-mm-long hairs. The pitchers of *H. folliculata* open bright yellow and suffuse orange and red as they age (Figure 100). Most plants consist of a beautiful array of leaves of different ages and colours.

Heliamphora glabra Nerz

> **Original description:** Nerz J., A. Wistuba and G. Hoogenstrijd, 2006,
> *Das Taublatt* 54: 58–70.

The specific epithet *glabra* is derived from the Latin *glaber* (hairless) and refers to the predominantly glabrous nature of the interior of the leaves of this species. *H. heterodoxa* var. *glabra* Maguire (1978) and *H. heterodoxa* fm. *glabra* (Maguire) Steyermark (1984) are taxonomically invalid and synonymous with *H. glabra*.

The range of *H. glabra* is currently known to include the summit of Wei Tepui (also spelled Uei and also called Serra da Sol) located southeast of Mount Roraima, and several extremely small, apparently unnamed cerro-plateaus on the borderlands of Brazil, Guyana and Venezuela (Figure 101). This species is also found on the summit of Wei Assipu (also called Little Roraima and Roraimita) and on the eastern flank of Mount Roraima. It is known to occur at several lowland locations near where the boundaries of Venezuela, Brazil and Guyana converge, including the swamp directly beneath "the prow," the northern tip of Mount Roraima. The altitudinal range of *H. glabra* is roughly 1200–2810 m above sea level. In many parts of its range, *H. glabra* grows and hybridizes with *H. nutans,* in some populations of which the hybrids are dominant.

Figure 101. The range of *Heliamphora glabra.*

H. glabra occurs in a wide variety of habitats ranging from open boggy marshlands, to dry, exposed ridge sides, to light *Bonnetia* cloud forests. In 0–20% shade, the leaves of *H. glabra* are 20–35 cm in length and extremely colourful. Each pitcher opens bright yellow, then slowly suffuses orange and, eventually, reddish purple as it ages. Usually each plant consists of a beautiful mixture of different aged and coloured leaves. In the 20–45% shaded conditions, the leaves grow up to 45 cm in length — among the largest leaves of all *Heliamphora* species — and are predominantly green with some red colouration. In more densely shaded conditions the foliage of *H. glabra* is pure green and etiolated. The colouration of the leaves of this species varies significantly between populations on different mountains. The *H. glabra* plants on Wei Tepui are predominantly green while those of the smaller mountains to the east of Mount Roraima display more vibrant red colouration.

The pitchers of *H. glabra* are 10–45 cm tall and 5–15 cm wide (Figure 102). The leaves of this species are proportionately narrower and more elongated than those of most other *Heliamphora* species. The lower section of the leaf is infundibular, but very close to the base — about one-third of the way up the pitcher — the girth is restricted which gives the lower portion of the leaf a slightly ventricose profile. The mid-section of the pitcher is tubular and slender, whereas the upper section is slightly infundibular towards the pitcher opening. The nectar-spoon lid is helmet shaped and has a distinctive knob-like bulge at its apex which is consistently present. This feature is also reliably inherited in *H. glabra x nutans* hybrids. The interior surface of the pitcher opening lacks downwards-pointing hairs, but a ring of 2–5-mm-long hairs is present above the water line on the interior surface of the leaf. The nectar of *H. glabra* is often infected by a black soot mould. Unlike most *Heliamphora* species, the floral bracts of this species are yellow rather than red.

Heliamphora heterodoxa Steyermark

Original description: Steyermark, J., 1951, *Fieldiana, Botany* 28: 239–242.

The specific epithet *heterodoxa* is derived from the Greek *heteros* (other) and *doxa* (opinion) and refers to the discovery of this as a new species of *Heliamphora* that was distinct from *H. nutans* and *H. minor,* the only other known species at the time that *H. heterodoxa* was discovered and named.

Figure 102. The leaves of *Heliamphora glabra*.

Two valid taxonomic varieties of *H. heterodoxa* are discernable; the type variety (*H. heterodoxa* var. *heterodoxa*) which occurs only on Ptari Tepui and a currently unnamed variety which makes up every other known population of *H. heterodoxa*, including all known lowland colonies of this species. The currently unnammed variety is not known to occur on Ptari Tepui. Overall, the unnamed variety is the more common of the

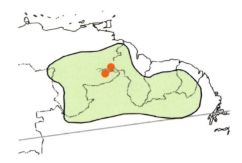

Figure 103. The range of *Heliamphora heterodoxa.*

two, despite the fact that it is not the type variety. *H. heterodoxa* (Steyermark) var. *exappendiculata* (Maguire) and *H. heterodoxa* var. *glabra* Maguire (1978) are invalid varieties of *H. heterodoxa* and have been re-named as separate species (see *H. exappendiculata* and *H. glabra* above). *H. heterodoxa* var. *exappendiculata* fm. *glabella* Steyermark (1984) and *H. heterodoxa* fm. *glabra* (Maguire) Steyermark (1984) are taxonomically invalid and synonymous with *H. glabra.*

The range of *H. heterodoxa* encompasses the summits of several plateaus within the Chimanta Massif as well as the distant and very isolated summit of Ptari Tepui in the northern part of the Gran Sabana (Figure 103). All of the tepui summits where highland populations of *H. heterodoxa* occur lie between 1900–2300 m above sea level. However, *H. heterodoxa* also occurs in wet patches of the lowland savannah be-tween 1100–1400 m above sea level from close to the base of Ptari Tepui eastward towards and possibly into Guyana (Figure 104). Considering the relatively wide range of *H. heterodoxa*, additional highland and low-land populations are likely to be discovered as the Chimanta Massif and surrounding mountains are more intensively explored. The foothills of these tepuis in particular represent refuges where as yet undiscovered populations of *H. heterodoxa* may occur.

H. heterodoxa is known to hybridize naturally with *H. sarracenioides*. *H. heterodoxa* grows in close proximity to *H. pulchella* on the Chimanta Massif, and hybrids between these two species may occur in this area of near sympatry.

The leaves of the type variety, *H. heterodoxa* var. *heterodoxa* from Ptari Tepui only, are infundibular in the lower section (Figure 105). The girth is constricted at around one-third of the height of the leaf causing the base of the leaf to appear slightly ventricose. The upper section of the pitcher is cylindrical and proportionately elongated, and is evenly

Figure 104. A cluster of lowland *Heliamphora heterodoxa* leaves growing amidst tall vegetation.

lined with 2–3-mm-long, downwards-pointing hairs on the interior surface. The pitcher opening is narrow and the nectar spoon is concave, proportionately large and consistently red. The leaves range between 15–30 cm in height. Developing foliage is typically bright yellow and suffuses dark maroon as it ages.

Figure 105. Two clusters of leaves of the type variety of *Heliamphora heterodoxa* var. *heterodoxa*. This variety is found only on Ptari Tepui.

The unnamed and much more widely distributed variety of *H. heterodoxa* varies from the type variety in that the pitchers are typically 15–25 cm tall and more voluminous in profile (Figure 106). The upper section of the leaf is infundibular and generally broader than in the type variety. The leaves are predominantly yellowish green and the nectar spoon is pure red. The aging pitchers tend not to turn pure maroon as they die back as is the case in most *Heliamphora* species. There are few discernable differences between the highland and lowland populations. The lowland populations of *H. heterodoxa* are more numerous and more widely distributed than all other lowland-dwelling species of *Heliamphora*. The lowland plants predominantly occur in open marshy habitat and grow nestled amidst short grasses and herbaceous plants. Some strains form dense clumps up to 120 cm across.

Figure 106. The common variant of *Heliamphora heterodoxa*, shown here, has not been described or named taxonomically at the infraspecific level.

Heliamphora hispida Wistuba and Nerz

Original description: Nerz, J., and A. Wistuba, 2000, *Carnivorous Plant Newsletter* 29(2):37–41.

The specific epithet *hispida* is derived from the Latin *hispidus* (bristly) and refers to the lining of coarse hairs that is present on the interior of the leaves of this species.

The range of *H. hispida* is restricted to the southern part of the Neblina Range on the border between Venezuela and Brazil (Figure 107). *H. hispida* is known to occur on Pico Phelps and the surrounding Cerro Neblina area between 1800–3014 m above sea level and may also grow in similar habitat which continues north through the inaccessible central valleys of the Neblina Range.

The pitchers of *H. hispida* are 15–25 cm tall and relatively stout in shape (Figure 108). The base of the leaf is infundibular, but the girth is

Figure 107. The range of *Heliamphora hispida*.

Figure 108. A small cluster of typical leaves of *Heliamphora hispida*.

constricted at the midsection so that the lower portion of the leaf is slightly ventricose. The upper half of the leaf is infundibular, especially towards the pitcher opening, which is 5–8 cm wide. The interior of the leaf is lined with 1–4-mm-long, coarse, downwards-pointing hairs, but this characteristic is variable and greatly reduced in some strains. In some cases, the exterior of the leaves of *H. hispida* are also lined with 0.5–1-mm-long hairs. The nectar spoon is relatively small, concave and pure red in colour. The remaining parts of the leaf are predominantly yellowish green and lined with variable red veins. *H. hispida* predominantly grows in open-canopy cloud forest or scrubland habitat in extremely humid, 0–40% shaded conditions. It readily forms small clumps of plants up to 80 cm in diameter and grows nestled amidst the leaf litter and mossy undergrowth. *H. hispida* competes efficiently with surrounding vegetation and the rosettes grow through short undergrowth.

Heliamphora ionasii Maguire

Original description: Maguire, B.,1978, *Memoirs of the New York Botanical Garden* 29: 36–62.

The specific epithet *ionasii* honours Jonah Boyan who co-discovered this species during an expedition led by Bassett Maguire. Roman Latin lacked the letter J, therefore "ionasii" represents the Latin version of Boyan's first name.

The range of *H. ionasii* is restricted to a single valley located between Ilu and Tramen tepuis in the eastern part of the Gran Sabana, at roughly 1800–2150 m above sea level (Figure 109). *H. ionasii* does not occur on the summits of either Ilu or Tramen tepuis. In several locations within the Ilu-Tramen Valley, *H. ionasii* grows in the company of *H. elongata* and hybrids of these two species are very common.

Figure 109. The range of *Heliamphora ionasii*.

Figure 110. A cluster of *Heliamphora ionasii* leaves.

The pitchers of *H. ionasii* are the most beautiful of all *Heliamphora* leaves (figures 110 and 111). The base of the leaf is infundibular, but the girth is constricted halfway up the leaf and gives the lower half of the pitcher a ventricose aspect. The upper parts of the leaf are strongly in-fundibular, especially towards the pitcher opening which is broad and flared. The interior of the pitcher opening is lined with 5–11-mm-long,

Figure 111. The magnificent leaves typical of *Heliamphora ionasii*.

downwards-pointing hairs which project from small bump-like swell-ings present on the inner surface of the pitcher (Figure 112). A second set of 0.1–0.5-mm-long hairs is present on the inner, uppermost 2–3 cm of the pitcher opening. The lid is concave and consistently dark red or purple. The size of the leaves depends upon the habitat in which the plant grows, but generally the leaves of *H. ionasii* are among the largest

Figure 112. Long downwards-pointing hairs are typical in the leaves of *Heliamphora ionasii.*

in the genus. A large proportion of *H. ionasii* plants grown in cultivation are actually either hybrids with *H. elongata* or artificial crossbreeds produced by horticulturists.

H. ionasii occurs in two distinct habitats. The densest populations occur in open, boggy clearings and glades within the cloud forest along the sides of the Ilu-Tramen Valley. High levels of rainfall combined with runoff from the tepui summits suppress the growth of the cloud forest trees and, in their absence, the ground is carpeted with a thick but low-growing, 30–50-cm-tall undergrowth of grasses, orchids and bromeliads amidst which *H. ionasii* grows. Nestled within the tangle of foliage in 0–30% shade, *H. ionasii* grows vigorously and develops spectacular colouration. The exterior of the leaves suffuses pinkish orange and develops subtle red veins. In contrast, the interior develops a striking pinkish-red colour and is often mottled with blotches of yellow and orange (Figure 113). Growing in full sunlight in the open clearings, the leaves of *H. ionasii* are 20–30 cm tall and 10–14 cm broad.

H. ionasii also occurs within the cloud forest on ridges, slope sides and around the edges of the clearings within the Ilu-Tramen Valley. Providing that the forest canopy is not closed and the stunted cloud forest trees are sparsely distributed, *H. ionasii* is able to grow on the forest floor amidst the mossy trunks and leafy undergrowth. In these dank conditions, it competes efficiently with the surrounding vegetation and produces

Figure 113. *Heliamphora ionasii* growing in full sunlight amidst dense but low-growing vegetation. Note the vivid colouration.

Figure 114. *Heliamphora ionasii* growing in shaded conditions within dense cloud-forest habitat. Note the predominantly green colouration.

especially large, predominantly green, but not etiolated, leaves up to 45 cm tall (Figure 114). *H. ionasii* grows well and produces flowers in 35–55% shaded conditions. However, in yet darker forests, growth is weak and populations generally die out. In 40% shade, the leaves of *H. ionasii* are 30–45 cm tall and 14–18 cm broad. The leaves are etiolated in 60% or more shade. Maguire reported curtains of *H. ionasii* growing directly on small cliff faces within the cloud forest around the common base of Ilu and Tramen tepuis. The differences between *H. ionasii* that grow in these two habitats are environmental and the appearance of the leaves would change if the growing conditions were altered. In both habitats, *H. ionasii* grows mainly in moist or saturated humus and decaying leaf litter.

Since their discovery, the tropical pitcher plants of Southeast Asia (*Nepenthes*) have been known to occasionally trap rodents naturally. On the discovery of *N. rajah* on Mount Kinabalu in Sabah, Malaysia (Figure 115), Spenser St. John (1862) reported:

> *We sat before the tent . . . observing one of our followers carrying water in a splendid specimen of the Nepenthes rajah, we desired him to bring it to us, and found that it held exactly four pint bottles. It was 19*

inches in circumference. We afterwards saw others apparently much larger, and Mr. Low, [Sir Hugh Low] while wandering in search of flowers, came upon one in which was a drowned rat.

It is misleading to imply that any carnivorous plant regularly or habitually preys on rodents or is specifically adapted to trapping any animal larger than insects. It would be more accurate to suggest that rodents are trapped by *Nepenthes* species very rarely and mainly through chance. Evidently, the trapping mechanisms of certain *Nepenthes* species, which evolved to attract insects, also invite the attention of certain

Figure 115. *Nepenthes rajah* on Mount Kinabalu, Sabah, Malaysia.

rodents and, in exceptional circumstances, lead to the trapping of mice and rats — albeit presumably through the same processes as regular insect prey are caught.

Considering the large size of the pitchers of many *Nepenthes* species, it is easy to understand how rodents are lured to, and occasionally fall into, the voluminous traps and drown. Indeed, it is not only the *Nepenthes* species with the very largest leaf wells which occasionally trap animal prey. During 2003, I observed an individual *N. hirsute,* a relatively small species of *Nepenthes,* in the Maliau Basin in Sabah, Borneo, and in one pitcher I discovered the body of a greyish-brown, 57-mm-long (including tail) mouse. I concluded that the pitcher plant had trapped the mouse because three conditions existed:

1. The trap, which was nestled amidst the leaf litter, was within both the habitat and realm of rodents.
2. The plant offered a form or forms of bait (presumably nectar, trapped dead insects within the leaves and standing water) that was attractive to rodents.
3. The pitcher was sufficiently large to both enable the rodent to fall inside and be retained until it either drowned in the digestive liquid or died of exhaustion.

Similar to the known rodent-catching species of *Nepenthes,* the traps of several cloud-forest-dwelling species of *Heliamphora* are also nestled in the leaf litter within both the habitat and realm of many known rodent species. The same forms of bait are also (inadvertently) present and usually occur in greater quantity. Lastly, the pitchers of several *Heliamphora* species, notably *H. ionasii* and *H. tatei* 'Cerro Aracamuni / Cerro Avispa Variant' are as large as 45 cm in height (many times more voluminous than the leaves of *N. hirsuta*) and it appears, could easily retain a small rodent.

While there is no known evidence of rodent carnivory in *Heliamphora,* there also have been no studies of the types of prey which these plants catch in the wild. On a theoretical level only, the same conditions exist to make possible the occasional trapping of rodent prey in certain *Heliamphora* species as has been documented in certain species of *Nepenthes.* Long-term studies will be required to establish the full range of prey which the remarkably giant *Heliamphora* trap.

Heliamphora macdonaldae Gleason

Original description: Gleason, H. A., 1931, *Bulletin of the Torrey Botanical Club* 58(6): 367–368.

The specific epithet *macdonaldae* is not explained in Gleason's type description, but it is believed to honour an American collector and field botanist Mrs. MacDonald. *H. macdonaldae* is synonymous with the invalid taxa *H. tatei* fm. *macdonaldae* (Gleason) Steyermark (1984) and *H. tatei* var. *macdonaldae* (Gleason) Maguire (1978).

The range of *H. macdonaldae* remains vague due to the extremely remote and inaccessible nature of the region in which the species occurs. Currently, it is known to occur on the summits and lower hills of Cerro Duida and Cerro Huachamacare in Amazonas State, Venezuela (Figure 116). The altitudinal range of *H. macdonaldae* roughly extends between 1500–2300 m above sea level. No natural hybrids have been recorded, but *H. macdonaldae* frequently grows with *H. tatei* and hybrids between these species may occur.

The pitchers of *H. macdonaldae* are extremely beautiful (Figure 117). The base of the leaf is infundibular, however the girth is constricted around one-third of the way up the leaf which causes the lower portion to appear slightly ventricose in shape. The upper section of the leaf is elongated and slightly infundibular. The leaf is typically 16–25 cm tall and 4–6 cm wide. The interior is glabrous apart from a thin row of 2–4-mm-long, inward pointing hairs that are present on the very rim of the pitcher opening (Figure 118). A ring of 4–6-mm-long hairs is present above the water line on the interior surface of the leaf where the girth of the leaf is constricted. The nectar spoon is conical and positioned low on the back of the leaf, in a manner very distinct from *H. tatei*. The most striking characteristic of this species is the spectacular colouration that

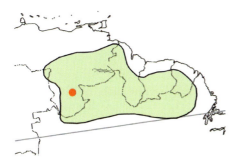

Figure 116. The range of *Heliamphora macdonaldae.*

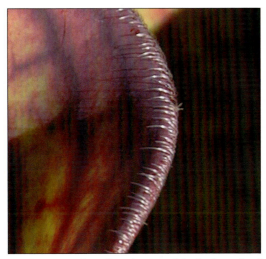

Figure 118. Inward-pointing hairs line the rim of the leaf opening in *Heliamphora macdonaldae.*

typically lines the interior of the leaf. The pitcher opening is lined with variable red or purple veins and suffuses pure red in some strains (Figure 119). This beautiful colouration is highly variable. The exterior of the leaf is consistently yellowish green and the nectar spoon is pure red. Old, established specimens of *H. macdonaldae* rarely form a short, decumbent stem, similar to that of *H. neblinae* and very different from that of *H. tatei*. Observations suggest that the population of *H. macdonaldae* on Cerro Huachamacare develop less vivid colouration than do other populations, and the foliage is predominantly yellowish green and suffuses red as it ages and dies back. While *H. macdonaldae* shares many characteristics with *H. tatei*, the leaves differ fundamentally in shape, and in particular, they lack the flared, very elongated trumpet shape of those of *H. tatei*. The nectar spoon in this species is not elongated and varies in shape from that of *H. tatei*. The interior of the leaf varies greatly from both *H. neblinae* and *H. tatei* in terms of its largely glabrous nature and unique colouration. Furthermore, the ecology and growing habits of *H. macdonaldae*, *H. neblinae* and *H. tatei* are very different.

Figure 117 (facing page). The beautiful, highly variable and uniquely coloured leaf of *Heliamphora macdonaldae.*

Figure 119. This strain of *Heliamphora macdonaldae* has leaves that suffuse pure red on the interior surface.

Heliamphora minor Gleason

> **Original description:** Gleason, H. A., and E. P. Killip, 1939, *Brittonia* 3(2): 141–204.

The specific epithet *minor* is derived from the Latin *minor* (smaller) and refers to the short leaves and compact growing habit of this species. *H. minor* fm. *laevis* Steyermark was described on the basis of an ecophene and is therefore taxanomically invalid.

The range of *H. minor* consists of the vast summit of Auyan Tepui in the northwestern part of the Gran Sabana (Figure 120). The species occurs between 1900–2500 m above sea level, but it is most prevalent at higher altitudes, especially on the southern half of the mountain. *H. minor* (Figure 121) is not sympatric with any other species of *Heliamphora* and consequently no hybrids occur naturally.

Figure 120. The range of *Heliamphora minor.*

The southern half of Auyan Tepui is desolate, barren and inhospitable. Up to 90% of the mountain surface is bare rock, scoured by the intense rain and washed clean of substrate and nutrients. In this cold, bleak setting, vegetation survives only where a roothold can be maintained, mainly in fractures and chasms in the rock surface or in the lee of

Figure 121. *Heliamphora minor* growing in full sunlight.

Figure 122. The growing habit of *Heliamphora minor*.

boulders and stones. Demand for the little available habitat is high. Rather than compete with one another, however, the plants of Auyan Tepui have evolved the tendency to grow together and form compact communities, collectively sheltering one another from the hardships of the climate. The short, stout nature of its leaves allows *H. minor* to grow nestled amidst dense but very short marsh grasses and orchids and to benefit from the protection which the other vegetation affords (Figure 122). Indeed, where *H. minor* grows alone, exposed to the elements, the size and vigour of those plants are greatly stunted. In very exposed areas, *H. minor* is restricted to gullies and hollows within the rock surface where it slowly divides and forms densely packed cushions of pitchers up to 80 cm across.

In more fertile areas, particularly the northern part of Auyan Tepui, *H. minor* readily grows amidst tall grass and shrubs. In 20–30% shade, the leaves are up to twice the typical size and predominantly green in colour (Figure 123). In 40% shade, the leaves are etiolated and *H. minor* populations generally die out.

The pitchers of *H. minor* are typically 10–15 cm in height and 3–5 cm wide at the pitcher opening. The base of the leaf is infundibular, the

Figure 123. Green leaves characterize this *Heliamphora minor* plant growing in 20% shade.

remaining parts of the leaf are stout and largely cylindrical. The girth of the pitcher is constricted only very slightly at the midsection and the upper section broadens slightly to the pitcher opening. The nectar spoon is generally helmet shaped, however in some populations it is flat and relatively broad. The interior of the leaf appears glabrous but is lined with microscopic 0.1–0.5-mm-long hairs. A narrow band of 1–3-mm-long, inward-pointing hairs is present at the very rim of the pitcher opening and a ring of 4–5-mm-long hairs is present on the interior of the midsection. The pitchers of *H. minor* do not have a drainage hole but rather a short, narrow drainage slit runs down the front of the leaf for 5–15 mm. In full sunlight, all parts of the leaves of *H. minor* are bright red.

Heliamphora neblinae Maguire

Original description: Maguire, B., 1978, *Memoirs of the New York Botanical Garden* 29: 36–62.

The specific epithet *neblinae* refers to Mount Neblina on the border of Venezuela and Brazil where this species occurs naturally. *H. tatei* var. *neblinae* (Maguire) Steyermark (1984), *H. neblinae* var. *parva* Maguire (1978), *H. tatei* var. *neblinae* fm. *parva* (Maguire) Steyermark (1984) and *H. neblinae* var. *viridis* Maguire (1978) are taxanomically invalid and synonymous with *H. neblinae*.

The range of *H. neblinae* (Figure 124) remains poorly known due to the extremely remote and inaccessible nature of the region in which it occurs (Figure 125). The type locality lies in the rain-fed valleys in the northwestern part of the Neblina Range, roughly at elevations between 1750–1850 m above sea level. Much confusion surrounds this species and most of the plants grown in cultivation under the name of *H. neblinae* are actually *H. tatei*. The *Heliamphora* populations distributed across the southern part of the Neblina Range, including Pico de Neblina, differ considerably from the type variety of *H. neblinae* and are most probably *H. tatei* or hybrids between *H. tatei* and *H. hispida*. At this time, the type form of *H. neblinae* is not known to occur in the southern part of the Neblina Range.

The summit of Mount Neblina is dominated by rolling upland meadows mixed with *Bonnetia* scrub and cloud forest. *H. neblinae* predominantly occurs amidst short grasses, bromeliads and orchids in a variety of habitats ranging from relatively dry, well drained ridges and slopes

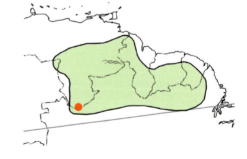

Figure 124. The range of *Heliamphora neblinae.*

Figure 125 (facing page). A population of *Heliamphora neblinae* living on the edge of oblivion.

to marshy bogs and the banks of small streams. Many aspects of the ecology of this species are distinctive, in particular its tendency to form a decumbent creeping stem up to 70 cm in length which enables it to scramble over rocks and colonize new habitat (Figure 126). In shaded locations *H. neblinae* occasionally grows upright on the length of its stem so that the leaf rosette is elevated above the surrounding vegetation and receives direct sunlight. However, this growing habit is extremely rare overall and, in most cases, upright *H. neblinae* are supported by surrounding vegetation and only grow 50–60 cm tall.

Approximately every 5 mm of stem length accommodates the growth of two leaves. Since each plant produces 6–12 pitchers annually, it is easy to estimate that large plants with stems longer than 50 cm are likely to be older than twenty years. The dead leaves remain attached to the stem of *H. neblinae* and efficiently collect and store rainwater which slowly leaks over the rhizome and roots and serves as a method of surviving drought. Fires regularly sweep across the Neblina Range and burn away the leaf litter which accumulates from the dense ground cover of grasses and shrubs. During these wild fires, the moist, dead pitchers which clothe the stems of *H. neblinae* protect the living core of the stem and enable most *H. neblinae* to survive the intensity of the heat. In the aftermath of a fire, the stems are charred and blacken but soon regrow and sprout fresh, green growth. Even if the apical bud is killed, new growth sprouts along the length of the stem or from the root system (Figure 127).

The pitchers of *H. neblinae* are typically 20–35 cm in height but can reach up to 40 cm in height in shaded conditions, and are 3–6 cm wide at the pitcher opening. The base of the leaf is infundibular, but the girth is constricted around one-third of the way up the leaf which causes the lower portion to appear slightly ventricose. The upper section of the leaf is infundibular but relatively narrow. The interior of the upper half of the leaf is lined with 1–3-mm-long, downwards-pointing hairs and a ring of 3–6-mm-long hairs is also present on the interior surface above the water line. The exterior of the leaves of *H. neblinae,* as well as the flower scapes and bracts, are consistently lined with 1–3-mm-long white hairs which, perhaps, are a means of protection against the intense high

Figure 126 (facing page). This *Heliamphora neblinae* plant shows the scrambling stem-forming habit of the species.

Figure 127. The main growth point of this *Heliamphora neblinae* plant was killed by an intense fire several months before this photograph was taken, but new growth is sprouting from the base of the stem.

altitude sunlight and fluctuating temperature. The nectar spoon of this species is highly variable and cannot be used for identification purposes. The pitchers of *H. neblinae* typically open yellowish green in colour and suffuse red with age, but some strains virtually lack red colouration altogether and are pure yellowish green (Figure 128). In other strains, the leaves have a distinctive glabrous red stripe that extends down the interior of the pitcher opening from the underside of the nectar spoon to about two-thirds of the way up the leaf. This trait is identical to that of the Cerro Aracamuni / Cerro Avispa variant of *H. tatei*, but is less pronounced and rarer in *H. neblinae* populations overall.

Nectar is secreted profusely from the nectar spoon and, once secreted, runs down the interior of the pitcher. An unidentified animal, perhaps a rodent, commonly rips off the nectar spoons of *H. neblinae* in order to eat the honey-sweet secretion. The lids are torn from the leaves and scattered on the ground around the plants, and they usually show signs that they have been chewed and scratched. *H. neblinae* also has established complex relationships with various species of ants. The ants

Figure 128. The leaves of *Heliamphora neblinae*, including (upper) those of typical individuals and (lower) those of a yellowish-green variant.

readily colonize the dead pitchers and use the shelter of the hollow leaves as nurseries for their larvae. In return for the shelter, the ants protect the green leaves of the plant they live in and swarm all over the plant if the leaves are disturbed. The ants are very rarely trapped by the plant and seem to regularly collect nectar from the nectar spoons.

H. neblinae produces larger flowers than any other species of *Heliamphora*. Each bloom is 50–80 mm in diameter and each tepal is 25–35 mm in width. Frequently, the flowers of *H. neblinae* are borne individually, although up to five blooms may be borne on a single scape. The number of tepals per flower is very unstable and frequently the flowers consist of 5 or 6 tepals rather than the usual 4. Flowers with differing numbers of tepals often occur on the same scape. The tepals fade reddish or greenish as they age. *H. neblinae* is most easily distinguished from *H. tatei* by the lining of 1–3-mm-long white hairs which consistently cover the exterior of the leaves of *H. neblinae* but are absent in *H. tatei*. Also, the morphology of the pitcher leaves of the two species differ profoundly. The pitchers of *H. tatei* are much more infundibular and elongated than those of *H. neblinae* and the nectar spoons differ entirely. The ecology of the two species is also very different.

Heliamphora nutans Bentham

> **Original description:** Bentham, G., 1840, *Transactions of the Linnean Society of London* 18: 429–432.

The specific epithet *nutans* is derived from the Latin *nutus* (nodding) and refers to the flowers of the species which move gently in the breeze. *H. nutans* was the first species of *Heliamphora* to be discovered. It was observed for the first time by Richard Schomburgk during an expedition to the base of Mount Roraima in 1838–1839. Schomburgk encountered it growing in a marshy swamp in the company of many completely new plant species. In "Journey from Fort San Joaquim, on the Rio Branco, to Roraima . . .," published in 1840, Schomburgk writes:

> *Another plant of great interest, the* Heliamphora nutans, *resembles the pitcher plant, which are similar to those of* Sarracenia variolaris; [*now named* S. minor] *but there was a great deviation in the flower; as in the present genus there are several flowers, and the seed are winged* (Schomburgk, 1840).

Forty years later, Sir Everard im Thurn led the first expedition to the summit of Mount Roraima, and encountered *H. nutans* growing on the plateau summit as well as in the lowland swamps.

> *In not very frequent places, where the grass is not so long, are considerable patches of the "pitcher plant" of South America* (Heliamphora nutans), *with its grotesquely pitcher-shaped leaves and delicate white flowers, borne on ruddy stems* (im Thurn, 1885).

The range of *H. nutans* consists of the summits of Mount Roraima, Kukenán Tepui, Wei Assipu Tepui (also called Little Roraima and Roraimita) and a number of extremely small, unnamed cerro-plateaus on the borderlands of Brazil, Guyana and Venezuela (Figure 129). On several occasions *H. nutans* has been reported to grow in the lowland swamps, including Schomburgk's "El Dorado" swamp, at the base of Mount Roraima, but many of these lowland populations have not been revisited and confirmed for several decades. The altitudinal range of this species lies roughly between 1200–2810 m above sea level. The *H. nutans* populations on Mount Roraima represent the most elevated *Heliamphora* except perhaps for isolated communities of *Heliamphora* on Pico Neblina. The temperature on the summit of Mount Roraima and Kukenán Tepui descends to just a few degrees above freezing each night and ranges between 15–25°C during the day, but, as is true of all of the tepuis, it is always free of ice or subzero temperatures. Curiously, *H. nutans* has a very uneven distribution on both Mount Roraima and Kukenán Tepui and occurs predominantly in marshy valleys around the edges of these mountains. In some areas it grows directly on rock overhangs on the cliff sides of these plateaus, and perhaps benefits from the ultra-humid ascending clouds which envelop the mountain sides. On Mount Roraima and Kukenán Tepui, *H. nutans* is strongly associated with other vegetation and grows mostly amidst short grasses and orchids (figures 130 and 131).

Figure 129. The range of *Heliamphora nutans.*

Figure 130. *Heliamphora nutans* growing amidst grasses and *Stegolepis.*

Figure 131 (facing page). *Heliamphora nutans* growing amidst dense, low-growing grasses and orchids.

Figure 132. The ornate leaves of *Heliamphora nutans*.

Where it does grow in more exposed habitat, the length of the leaves is reduced to 3–5 cm. The largest and most vigorous plants grow in 15–30% shade in well drained habitat.

The pitchers of *H. nutans* are typically 10–30 cm in length and 3–7 cm wide at the pitcher opening (Figure 132). The base and upper sections of the leaf are infundibular, but the girth is constricted at the midsection which causes the lower half of the leaf to appear slightly ventricose in profile. The leaves are predominantly yellowish green in colour and suffuse orange and red as they age. The nectar spoon is consistently red or purple and helmet shaped. The interior of the upper half of the leaf is lined with 1–3-mm-long, downwards-pointing hairs and a ring of 2–5-mm-long hairs is present above the water line. In the wild, *H. nutans* is generally much smaller than plants grown in cultivation. The overwhelming majority of *H. nutans* plants grown in cultivation are actually either hybrids with *H. glabra* or artificial crossbreeds produced by horticulturists.

Heliamphora pulchella Wistuba, Carow, Harbarth and Nerz

Original description: Wistuba, A., T. Carow, P. Harbarth and J. Nerz, 2005, *Das Taublatt* 53: 3.

The specific epithet *pulchella* is derived from the Latin *pulcher* (beautiful) and refers to the attractive leaves of this species. *H. pulchella* was previously referred to as *H. minor* 'Hairy Variety' or 'Chimanta Variety.'

The range of *H. pulchella* encompasses the tepui summits and foothills of the western part of the Gran Sabana (Figure 133). *H. pulchella* is extremely common across the Chimanta Massif as far north as Araopan Tepui and predominantly occurs between 1700–2400 m above sea level. *H. pulchella* is not recorded from the lowlands, but the inaccessible swamps at the base of the Chimanta Massif may hold lowland populations of this species. *H. pulchella* is known to hybridize with *H. chimantensis* but grows in close proximity to *H. heterodoxa* in several locations and hybrids between these two species may exist.

H. pulchella occurs in a wide variety of habitats ranging from shallow marshy ponds to the margins of the *Bonnetia* cloud forest (Figure 132). While it readily grows in open clearings away from surrounding vegetation, it also occurs very frequently nestled amidst short growing grasses and bromeliads in 0–30% shaded conditions. The dwarfed ecophene of *H. pulchella* is extremely common in dry and exposed habitat.

The pitchers are typically 8–12 cm tall and 3–5 cm wide at the pitcher opening and similar in shape to *H. minor* (Figure 134). The lower part, close to the base of the leaf, is infundibular. The girth is constricted only very slightly at the midsection and, depending upon the degree of shade, the upper parts are infundibular (in shaded habitat, the leaves are strongly infundibular). The small, helmet-shaped nectar spoon is slightly inset into the back of the leaf and is typically purple in colour. No drainage

Figure 133. The range of *Heliamphora pulchella*.

hole is present on the leaves of *H. pulchella;* instead, a 10–20-mm-long drainage slit runs down the front of the leaves.

In the typical variety, the interior of the middle and upper sections of the leaf is lined with very prominent, white, 3–7-mm-long hairs and also (usually) a much denser set of 0.1–0.5-mm long hairs (figures 135, 136 and 137). The two types of hairs occur together and dominate most parts of the interior surface of the leaf down to the midsection of the pitcher. A set of 1–2-mm-long hairs is also present on the inner surface of the midsection of the leaf and forms a dense white band just above the water line. All three sets of hairs terminate at the midsection. In some rare strains of *H. pulchella* either one or two of these sets of hairs, in any combination, can be altogether absent, including the distinctive long hairs (figures 74 and 138).

Figure 135. The typical variant of *Heliamphora pulchella* growing in full sunlight. Note the prominent hairs.

Figure 134 (facing page). The typical variant of *Heliamphora pulchella* growing in marshy habitat.

Figure 136. The typical variant of *Heliamphora pulchella* growing in 20% shade. Note the prominent hairs.

The foliage of *H. pulchella* is typically pure red or purple in full sunlight, but the colouration of the leaf responds very dramatically to the degree of light available and in 20–40% shade, the leaves are predominantly green with variable reddish colouration and veins (Figure 136). In localized populations, the colouration of the pitchers is especially dark and the aging leaves turn a deep shade of purplish black (figures 74, 137 and 138). This colouration contrasts vividly with the younger leaves which open bright scarlet in colour and slowly suffuse maroon and burgundy. It is not known whether this colouration trait is genetic or the result of the local growing conditions, although it is probably a combination of the two. The most intensely coloured *H. pulchella* always grow exposed to direct sunlight, usually in very wet or semi-aquatic conditions. Perhaps the light reflected from the water surface increases the ambient light levels and darkens the colouration of the leaves.

The differing shape, colouration, stature and typically long-haired nature of *H. pulchella* help to distinguish it from *H. minor*.

Figure 137 (facing page). A cluster of leaves of the purple black variant of *Heliamphora pulchella*.

Figure 138. A deep purple black variant of *Heliamphora pulchella* that lacks prominent downwards-pointing hairs.

Heliamphora sarracenioides Carow Wistuba and Harbarth

Original description: Carow, T., A. Wistuba and P. Harbarth, 2005, *Carnivorous Plant Newsletter* 34(1): 4–6.

The specific epithet *sarracenioides* refers to the similarities between this species and those in the genus *Sarracenia*.

The range of *H. sarracenioides* consists of the summit of a single undisclosed mountain in the northern part of the Gran Sabana at altitudes of 2500–2650 m above sea level. *H. sarracenioides* is known to naturally hybridize only with *H. heterodoxa*. It predominantly grows amidst short marsh grasses in full sunlight, in extremely wet and humid conditions.

The pitchers of *H. sarracenioides* are distinctive (Figure 139). The leaf is 10–20 cm tall and 2–4 cm wide at the pitcher opening. The lower parts of the leaves are slightly ventricose while the upper and middle sections are cylindrical or slightly tapered towards the pitcher opening. The girth of the leaf is constricted only very slightly at the midsection of the leaf. Overall, the leaf of *H. sarracenioides* is much more tubular in shape than in all other species of *Heliamphora* and consequently appears more similar to the *Sarracenia* species. Like the North American pitcher

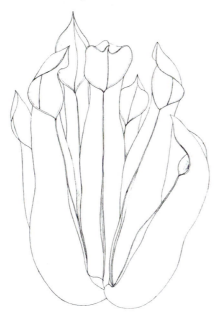

Figure 139. The mature leaves of *Heliamphora sarracenioides*.

plants, *H. sarracenioides* lacks a *Heliamphora*-type nectar spoon and, instead, the back of the leaf extends over the pitcher opening and forms a hood-like covering similar to what is found in many species of *Sarracenia*. The hood shelters the pitcher from rain, although a drainage hole is present on the leaves of *H. sarracenioides*. The interior of the pitcher opening is entirely glabrous apart from a ring of 2–5-mm-long, downwards-pointing hairs present on the interior surface of the leaf, just above the water line. Nectar glands are distributed across the interior of the hood. The leaves of *H. sarracenioides* are either pure red or reddish green in strong sunlight. The interior of the hood is dark red or purple. It remains unclear what *H. sarracenioides* represents. Future DNA studies will shed light on whether this species represents an ancient link with the *Sarracenia* or perhaps just a recent mutation.

Heliamphora tatei Gleason

> **Original description:** Gleason, H. A., 1931, *Bulletin of the Torrey Botanical Club* 58(6): 367–368.

The specific epithet *tatei* honours George Tate of the American Museum of Natural History who led various expeditions to the Guiana Highlands during the 1930s. *H. tatei* var. *neblinae* (Maguire) Steyermark (1984) is an invalid synonym of *H. neblinae* Maguire. In 1931, Gleason published three species (*H. macdonaldae*, *H. tatei* and *H. tyleri*) which he observed during his expedition to the summit of Mount Duida. The type varieties of *H. tatei* and *H. macdonaldae* are distinct from each other and certainly can be considered valid species. The type variety of *H. tyleri* has not been recorded in the wild since its discovery and, at this time, it is believed to be synonymous with *H. tatei*. The only significant difference between *H. tyleri* and *H. tatei*, which Gleason (1931) notes, is that the shape of the lid of *H. tyleri* differs and is "broadly obovate" (Figure 140) — this does not warrant granting *H. tyleri* status as a separate species, especially since this variant appears to be very similar to the Cerro Aracamuni / Cerro Avispa variant of *H. tatei* and is well within

Figure 140 (facing page). The unique growing habit of *Heliamphora tatei*. This variant would have been referred to *Heliamphora tyleri* by Gleason due to the shape of the nectar spoon — it is not conical.

the natural range of variation of *H. tatei*. *H. tatei* var. *macdonaldae* (Gleason) Maguire (1978) and *H. tatei* fm. *macdonaldae* (Gleason) Steyermark (1984) are invalid synonyms of *H. macdonaldae* (see above). *H. tatei* var. *neblinae* (Maguire) Steyermark (1984) and *H. tatei* var. *neblinae* fm. *parva* (Maguire) Steyermark (1984) are also invalid (see *H. neblinae* above).

The range of *H. tatei* remains poorly known due to the extremely remote and inaccessible nature of the region in which it occurs. *H. tatei* was first discovered on the summits and foothills of Cerro Duida and Cerro Huachamacare between 1700–2400 m above sea level, but it is also known to occur at similar altitudes on the side hills of Cerro Marahuaca and, possibly, in the southern part of the Neblina Range (Figure 141). Populations of this species also are known from Cerro Aracamuni and Cerro Avispa, but the plants on these mountains differ slightly from the type variety (see below). No hybrids of *H. tatei* have been recorded in the wild, but on Cerro Duida and Cerro Huachamacare *H. tatei* frequently grows in the company of *H. macdonaldae* and hybrids between these two species may occur.

H. tatei has evolved a unique strategy to compete with the dense vegetation of the marshy grasslands, heath scrublands and sparse *Bonnetia* cloud forest amidst which it grows. It forms an erect, woody stem which supports and elevates the leaf rosette up to 2 m into the air above the surrounding undergrowth, thereby better positioning the leaves to photosynthesize and attract prey. When the stem grows longer than 2 m, the weight of the water-filled leaves usually causes the stem to bend over. In most cases, the bent *H. tatei* continues to grow and the angle and direction of the stem adjusts so that the rosette grows upright. In this way, *H. tatei* can form curved, snaking stems up to 4 m in length. Similar to *H. neblinae*, the stem of *H. tatei* is an accurate record of the age of a particular

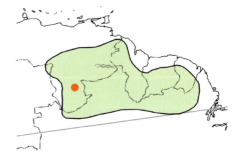

Figure 141. The range of *Heliamphora tatei*.

individual. The stem of each plant grows 4–10 cm in length annually so it is easy to estimate that large plants with long stems are likely to be several decades old.

The pitchers of *H. tatei* are 25–35 cm in length, the lower section of the leaf is slightly infundibular, however narrow and proportionately elongated. The girth of the leaf is constricted at the midsection and causes the lower half of the leaf to appear very slightly ventricose. The upper half of the leaf is infundibular, especially towards the pitcher opening which is typically 5–9 cm broad. A drainage hole is present one-third of the way up, on the front of the leaf and usually, although not always, a 10–40-mm-long V-shaped slit extends down the front of the leaf from the pitcher opening. More than any other species of *Heliamphora*, *H. tatei* depends upon these drainage mechanisms to regulate the level of rain water and digestive fluids contained within its leaves. If the pitchers were to fill up with water entirely, the weight would cause the long, cane-like stems to bend over or collapse. Also for this reason, each plant almost always consists of four or fewer living, water-filled leaves.

The nectar spoon of the type form of *H. tatei* is small and generally conical in shape (Figure 142). The back wall of the pitcher is narrow, elongated and often slightly slanted towards the front of the leaf. The interior of the upper half of the leaf is lined with 1–3-mm-long, downwards-pointing hairs, and a ring of 4–8-mm-long hairs is present just above the water line on the interior of the leaf. *H. tatei* is highly variable in terms of colouration and it remains unclear which variety is most representative of the species. In some populations the young opening leaves are bright yellowish white in colour and slowly darken to pure apple green as they age. Elsewhere, the pitchers are predominantly yellowish green with pure red nectar spoons. Other strains have prominent red veins or a glabrous red stripe that runs down the interior of the back wall from the nectar spoon. The pitchers of all known varieties of *H. tatei* usually suffuse pure orange, red or scarlet as they age and die back.

Unlike the stems of *H. neblinae*, the dead leaves soon break away from the stems of *H. tatei* so that 50 cm beneath the leaf rosettes, the stems are naked and cane-like. This tendency perhaps evolved to reduce wind resistance by the stem or to decrease the weight which the stem has to support. The stems of *H. tatei* never divide along their length as this would imbalance the vertical structure and render it likely to fall over. Instead, *H. tatei* has a limited ability to produce offshoots from the

Figure 142. The leaf of the type form of *Heliamphora tatei* showing clearly the conical-shaped nectar spoon.

very base of its stem at ground level, but most often each plant consists of a single long stem and a single leaf rosette. The pitchers of *H. tatei* are larger, more infundibular and more elongated than the leaves of *H. neblinae*. The differences of the flowers, the nectar spoons and the stems also help to differentiate the two species. In the wild, the different growing habits are also obvious: *H. neblinae* almost always grows horizontally, scrambling along the ground on the length of its stem while *H. tatei* almost always grows erect, or snakes through the undergrowth, on far longer stems.

The populations of *H. tatei* on Cerro Aracamuni and Cerro Avispa differ from the typical variety of the species in many ways (Figure 143). They display a clear preference for densely vegetated habitat and grow most prevalently in 20–35% shaded conditions amidst thick undergrowth

Figure 143. The leaves of the Aracamuni-Avispa variant of *Heliamphora tatei*. Note the difference in the shape of the nectar spoon between this variant and the type form.

of grasses, orchids, bromeliads and small shrubs. The leaf structure is similar to the typical variety of *H. tatei,* but the nectar spoon is not conical and the colouration is extremely uniform. The leaves are pure green and marked with a distinctive glabrous red stripe that extends down the interior from the nectar spoon to half way down the leaf. The variant found on Cerro Aracamuni and Cerro Avispa does not form an erect stem but rather grows terrestrially, like the majority of *Heliamphora* species. Where the surrounding undergrowth is thick and encroaches to such a degree as to hinder growth, this variant displays a restricted ability to grow upright through vegetation and its stem seldom if ever exceeds 30 cm in height.

The Natural *Heliamphora* Hybrids

The maps in this chapter illustrate that in many areas of the Guiana Highlands the ranges of multiple species of *Heliamphora* overlap so that several species occur on the same tepuis. In these areas the same insects, mainly wasps and possibly beetles, visit the flowers of all species of *Heliamphora* present and often cross-fertilize the flowers by transferring the pollen of one species to the flowers of another. The resulting offspring inherit an intermediate blend of the characteristics of both parent species and is called a crossbreed, or hybrid.

Many species of *Heliamphora* grow in isolation so not all of the possible combinations of *Heliamphora* hybrids occur naturally. However, where different species do grow together, hybrids generally are very common and, in some cases, more numerous than individuals representing pure species. Since all *Heliamphora* are inter-fertile, hybrids can reproduce with other hybrids or species and a large number of different combinations might result. Currently, five *Heliamphora* hybrids have been observed in the wild, but it is likely that many more actually occur.

Heliamphora chimantensis x pulchella

On the summits of various tepuis of the Chimanta Massif, *H. chimantensis* and *H. pulchella* frequently grow together and readily hybridize. The crossbreed between these two species produces leaves that are 10–15 cm tall and similar in shape to those of *H. pulchella* (Figure 144). The pitchers are predominantly orangey pink or red in colour and develop delicate purple veins. Similar to *H. chimantensis*, this hybrid forms large clumps up to 2 m in diameter.

Heliamphora elongata x ionasii

The *Heliamphora elongata x ionasii* hybrid is a particularly beautiful blend of the excellent colouration and form of its two parent species. The leaves are predominantly red, pink and orange and usually lined with subtle purple veins (Figure 145). Typically, the pitchers are 20–35 cm in height, 10–15 cm wide and sturdy in shape (similar to *H. elongata*) yet infundibular and voluminous (similar to *H. ionasii*). *H. ionasii* and *H. elongata* only grow together within marshy clearings in the Ilu-Tramen Valley, and locally within this area of sympatry their hybrids occur very frequently and in some populations are more prevalent than

Figure 144. The leaves of *Heliamphora chimantensis x pulchella* hybrid.

pure individuals of either species. Many of the plants grown in cultivation as *H. ionasii* are actually *H. elongata x ionasii* hybrids. The leaves of *H. elongata* x *ionasii* are lined with 2–4-mm-long hairs and lack the distinctive bump-like swelling on the interior surface of mature *H. ionasii*.

Figure 145. The leaves of *Heliamphora elongata x ionasii* hybrid.

Heliamphora exappendiculata x glabra

This hybrid is known only from a handful of localities across the western tepuis where *H. exappendiculata* and *H. glabra* grow together or in close proximity. The pitchers are 15–25 cm tall and 7–11 cm wide and are predominantly yellowish green in colour but suffuse red as they age. The nectar spoon is small and helmet shaped but fused to the back of the pitcher, reflecting a combination of characteristics of the parent species. The interior of the leaf is lined with 1–3-mm-long downwards-pointing hairs much like *H. exappendiculata*.

Heliamphora glabra x nutans

On the summit of Wei Assipu Tepui (also called Little Roraima and Roraimita) and several of the unnamed cerro-plateaus located on the borderlands of Brazil, Venezuela and Guyana, *H. glabra* and *H. nutans* grow together and hybridize freely. Crossbreeds are often prevalent and frequently out-number both of the parent species.

In bright habitat, the leaves of *H. glabra x nutans* are 15–25 cm in height and 4–8 cm wide. However, where the hybrid grows within sparse cloud forest and in more shaded conditions, it produces immense leaves up to 45 cm tall. In sunny habitat, the leaves of *H. glabra x nutans* develop beautiful orangey-red colouration while plants growing in densely shaded conditions are predominately greenish yellow. The nectar spoon in this species is consistently red and always bears the characteristic knob-shaped apex of *H. glabra*. Interestingly, the interior of the leaves of *H. glabra x nutans* are usually only partially lined with downwards-pointing hairs. A hairless patch is generally present on the back of the interior of the pitcher and this reflects the conflict of characteristics between *H. nutans* (leaves lined with hairs) and *H. glabra* (largely glabrous leaves).

H. glabra x nutans readily re-hybridizes with both parent species and complex backcrosses are relatively common. On several occasions, *H. glabra x nutans* plants have been mistaken for pure *H. nutans* and many of the "giant" forms of *H. nutans* that are grown in cultivation are actually this hybrid.

Heliamphora hispida hybrids

H. hispida grows in the southern part of the Neblina Range, as does *H. tatei* and possibly another undescribed taxon. However, since the *Heliamphora* populations in this region appear to be a large and complicated hybrid swarm, it is difficult to identify with certainty the taxonomic identity or lineages of individuals and populations. Many of the plants grown as *H. neblinae* in cultivation are certainly hybrids involving *H. hispida*. Hybrids of the southern part of the Neblina Range are extremely difficult to identify and probably represent the result of several generations of crossbreeding.

Sarracenia

The genus *Sarracenia* — and family Sarraceniaceae — were named in honour of the Canadian physician Dr. Michel Sarrazin who, at the start of the 18th Century, secured twenty-five specimens of pitcher plants, including *S. purpurea,* and sent them to the French botanist Joseph Pitton de Tournefort for classification. In gratitude, Tournefort later named the genus *Sarracenia* in honour of Sarrazin.

Sarracenia (Figure 146) was the first genus of pitcher plants discovered in the New World. The earliest known record is an illustration and brief mention of *S. minor* as *Thuris limpidifolium* (Figure 147) in Matthias de l'Obel and Petrus Pena's *Nova Stirpium Adversaria,* first printed in 1570 and again, in the Dutch version *Kruydtboeck . . .,* published in 1581. In these works, l'Obel records that he received two leaves of *Thuris limpidifolium* via a French doctor and was informed that they originated from a resinous pine tree. The next known reference was a more formal description accompanied by an accurate illustration of *S. purpurea* as *"Limonio congene"* in Carolus Clusius's *Rariorum Plantarum Historia,* published in 1601. Clusius was confused and amazed by the hollow structure of the leaves and was not sure of their purpose. During the 17th and 18th centuries, several pioneering botanists further documented *Sarracenia* species. Mark Catesby, one of these early botanists, described and depicted *S. flava, S. minor* and *S. purpurea* in detail under different names in *The Natural History of Carolina, Florida, and the Bahamas,* published in 1754 (figures 148 and 149). Catesby noted that "The hollows of these leaves . . . always retain some water, and seem to serve as an asylum or secure retreat for numerous insects."

Figure 146 (facing page). A mixture of *Sarracenia* species flowering in the Gulf Coast region of the United States.

Figure 147. *Thuris limpidifolium* featured in *Nova Stirpium Adversaria* (1570).

The notion that the tubular shape of the leaves may be an evolutionary modification that enabled carnivory emerged in the 19th Century following a report written by James MacBride and published in 1815 in *Philosophical Transactions of the Royal Society*. MacBride conducted a series of detailed observations and wrote of the trapping processes of *Sarracenia* species.

> *It will hardly be necessary to inform you that the* Sarracenia flava *and* S. adunca . . . *grow in the flat country of this state in great adundance. With the latter my experiments have been chiefly conducted. If . . . the leaves of these plants . . . be removed to a house and fixed in an erect position, it will soon be perceived that flies are attracted by them. These insects immediately approach the fauces of the leaves, and leaning over their edges appear to sip with eagerness something from their internal surfaces. In this position they linger; but at length, allured as it would seem by the pleasure of taste, they enter the tubes. The fly which has thus changed its situation, will be seen to stand unsteadily, it totters for a few seconds slips, and falls to the bottom of the tube, where it is either drowned, or attempts in vain to ascend against the points of the hairs . . . In a house much infested by flies, this entrapment goes on so rapidly that the tube is filled in a few hours, and it becomes necessary to add water, the natural quantity being insufficient to drown the imprisoned insects. The leaves of the* S. adunca *and* S. rubra *of Walter*

Figure 148. *Sarracenia minor* and *Sarracenia flava* as presented in Catesby's *The Natural History of Carolina, Florida and the Bahama Islands* (1754).

might well be employed as flycatchers; indeed I am credibly informed they are in some neighbourhoods. The leaves of the flava, *although they are very capacious and often grow to the height of three feet or more, are never found to contain so many insects as the leaves of the species above mentioned. The spreading fauces and erect appendices of the leaves of this species, render them (I suppose) less destructive. The cause which attracts flies is evidently a sweet viscid substance, resembling honey, secreted by, or exuding from, the internal surface of the tube.* (MacBride, 1815; MacBride notes that *S. adunca* Smith is an invalid synonym of *S. minor* Walter.)

In his book *Insectivorous Plants,* published in 1875, Charles Darwin documented the results of various sets of meticulous experiments which he had conducted to prove conclusively that "There is a class of

Figure 149. *Sarracenia purpurea* as presented in Catesby's *The Natural History of Carolina, Florida and the Bahama Islands* (1754).

plants which digest and afterwards absorb animal matter." At his home in Kent, England, Darwin grew several species of *Sarracenia* and corresponded with various American and British naturalists from whom he acquired various specimens and information. While he did not have the opportunity to study *Sarracenia* extensively, during the later years of his life Darwin developed firm beliefs that "There can scarcely be a doubt that *Sarracenia* ... may be added to this class (i.e., carnivorous plants), though the fact can hardly be considered as yet fully proved" (Darwin, 1875).

During 1874 and 1875, Dr. J. H. Mellichamp and Dr. W. M. Canby conducted a series of observations on *Sarracenia* species in order to better understand the possible carnivorous nature of that genus. They focused their efforts on examining herbarium specimens of *S. minor* collected by Mellichamp in South Carolina during the early 1870s and eventually concluded "that the fluid of the pitchers [of *Sarracenia* sp.] hastens the decomposition of insects" (Lloyd, 1942) and apparently enable the pitcher plants to "digest" animal prey. In 1885, Professor William Kerr Higley of The Chicago Academy of Sciences published a study of the liquid contained within the leaves of *S. purpurea* and demonstrated that ammonia and nitrogenous compounds were absorbed by the leaves of *Sarracenia* plants. However, according to Lindquist (1975), it was the studies of Fenner (1904) that finally proved the carnivorous nature of the *Sarracenia*.

Fenner, 1904 added flies to the lowermost part of a pitcher of Sarracenia flava *without added water and noticed a mucilaginous secretion pouring out of the pitcher lining and digesting the insect bodies in contact with the lining in the course of several hours. The mass of insects became saturated with the fluid, and putrefaction was entirely absent until this "absorptive" zone was filled and additional insects began collecting in the hairy "detentive" zone. Thus* S. flava *was shown to be a truly insectivorous plant with a digestive enzyme* (Lindquist, 1975).

Plant Structure

The *Sarracenia* are clump-forming, herbaceous perennials that produce hollow cylindrical or tubular leaves of varying sizes and structures. The foliage emerges from a subterranean branching rhizome and, in most species, is arranged in a circular rosette. Fibrous brown roots 20–30 cm long emanate along the length of the rhizome. In all *Sarracenia* species, the flowers are borne individually on 15–70-cm-tall scapes and are structurally consistent throughout the genus. The most distinguishing feature of the *Sarracenia* bloom is the upturned umbrella-shaped style over which the five petals loosely hang. The edge of the style undulates five times and the five stigmas face inwards on the peaks of the undulations. At the centre of the flower, at the base of the style, is the five-chambered ovary. The 20–60 stamens are arranged around the ovary beneath the five strap-shaped or rounded petals. The anthers are oblong, 6–10 mm in length and yellow in colour. The peduncle is glabrous. Three small rounded bracts are present just above the five lanceolate sepals (Figure 150).

The growth of *Sarracenia* follows an annual cycle determined by the seasons. For 2–3 weeks during April and May, each species flowers and then produces carnivorous leaves over the course of late spring and

Figure 150. A bee visiting the flower of *Sarracenia purpurea* ssp. *venosa* var. *burkii*.

summer before becoming dormant in autumn. Within 1–2 days after the flowers open, the stigmas become receptive and, simultaneously, the anthers shed their pollen. The petals and stamens die and fall to the ground 6–10 days later. If, during that time, the flower has been pollinated, the ovary starts to swell and over the course of the summer it develops between 20–300 seeds which ripen and are distributed during autumn. *Sarracenia* seeds are pear shaped and yellow, light brown or purplish in colour. Bees and winged insects are the main pollinators of *Sarracenia* flowers and, since many *Sarracenia* species often grow closely together, hybridization frequently occurs. The shape and colouration of the flower varies among the various species and varieties.

During the first year following germination, young *Sarracenia* plants produce simple tubular juvenile leaves which are largely the same in all species and reminiscent of the mature leaves of *S. minor*. The juvenile leaves are generally 1–3 cm in length and reddish in colour. During the second, or occasionally third year of growth, *Sarracenia* species spontaneously begin to produce adult leaves, and they finally reach full maturity and flower 3–6 years after germination. The shape and colouration of the adult carnivorous leaves vary among the eight *Sarracenia* species but share a common structure (figures 151 and 152). In all species, the upper section of the leaf serves the purpose of attracting insects through vivid colouration, often red or purple which contrasts with grassland habitat, and copious nectar-secreting glands which produce droplets of nectar and create a strong sweet scent.

In all eight species, the interior of the leaf is smooth, waxy and slippery, especially at the opening of the pitcher. Throughout the genus, the apex of the leaf forms a "lid" which shelters the tubular portion of the trap and prevents it from filling with water and toppling over. The leaves of *Sarracenia* are generally between one-sixth and one-third filled with water and secreted digestive liquids. The interior surface of the very bottom of the leaf is sparsely lined with 2–5-mm-long hairs.

In addition to carnivorous leaves, *S. alata*, *S. flava*, *S. leucophylla* and *S. oreophila* also produce ensiform foliage adapted specifically for photosynthesis.

Trapping Process

S. alata, *S. flava*, *S. leucophylla*, *S. oreophila*, *S. purpurea* and *S. rubra* employ essentially the same trapping process, one that is strikingly similar to that of the *Heliamphora* species. Insects are attracted to the leaves by the conspicuous colouration and profuse secretions of nectar concentrated around the pitcher opening (Figure 153). The leaves have a distinctly sweet scent which attracts winged insects very effectively. Ground insects follow nectar trails up the exterior of the pitcher leaf, and these trails strengthen towards the entrance of the trap. Once attracted, insects detect that the heaviest concentrations of nectar are present on the back of the leaf beneath the lid, and this forms the bait in the trap. The inner surface of the pitcher opening is waxy and very slippery, however to reach the nectar the insect has to precariously scale the interior of the leaf and hang on to the vertical back of the leaf directly over the tubular trap (Figure 153). A foothold is very difficult for the insect to maintain, and the slightest falter causes it to plummet straight down into the depths of the leaf and into the digestive liquid contained within. Once inside the trap, the unfortunate victim is unlikely to escape as it is unable to scale the waxy interior surface of the leaf, and eventually it drowns or dies of exhaustion. The remains of trapped prey accumulate at the bottom of the leaf.

Figure 151. A simplified *Sarracenia* plant, with (A) scape, (B) flower, (C) seed pod, (D) seed and (E) seedling also shown.

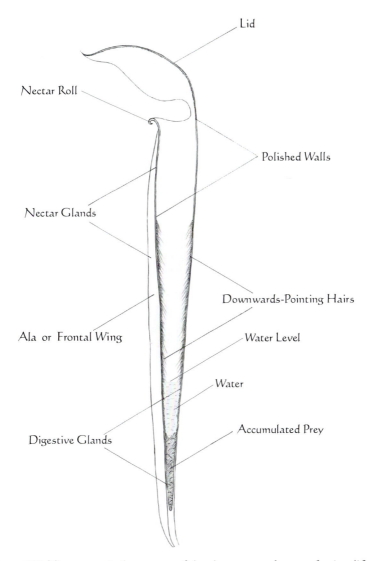

Lid

Nectar Roll

Polished Walls

Nectar Glands

Downwards-Pointing Hairs

Ala or Frontal Wing

Water Level

Water

Accumulated Prey

Digestive Glands

Figure 152. The anatomical structure of the above-ground parts of a simplified *Sarracenia* plant, with pitcher contents added.

The trapping processes of *S. minor* and *S. psittacina* differ from those of all other *Sarracenia* species and share obvious similarities with *D. californica*. In *S. minor,* nectar is secreted mainly at the entrance of the pitcher opening, beneath the long hood-like lid. The back of the leaf is lined with large, white, semi-translucent areoles which are brightly

Figure 153. A hover fly, attracted to a leaf of *Sarracenia flava*, is (left) lured to the nectar bait where (right) it positions itself precariously over the tubular trap.

illuminated in sunlight (Figure 154). As in the other species of *Sarracenia*, insects are attracted to the leaves by their bright colouration and sweet scent and, once at the plant, crawl underneath the hood to reach the nectar concentrations at the entrance of the trap. The long lid overhangs and shades the pitcher opening so that, from the position of the insect, the white areoles at the back of the leaf present the brightest route and apparently the direction of bright sky and freedom. After eating the nectar, the insect flies directly towards the areoles, collides with the back of the leaf and falls straight down into the depths of the trap where it drowns and is eventually digested.

In the case of *S. psittacina*, a similar trapping process is employed although in this species the leaves are arranged horizontally and rest directly on the surface of the ground, often in very wet habitat (Figure 155). The entrance of the trap is a small, inward-protruding, funnel-shaped hole present on the centreward side of the dome-shaped hood (Figure 156). The scent of nectar inside the leaf encourages insects to venture to the entrance of the trap where they find the secretions. Similar to *S. minor*, the overhanging shape of the hood, in this case modified to form a dome structure, causes the sunlit fenestration at the back of the pitcher to represent the brightest route which the insect misinterprets as bright sky and the way out of the leaf. It enters the trap and flies directly towards the areoles but, naturally, collides with the back of the leaf and falls to the bottom of the dome. Like the entrance of a lobster pot, the inward-protruding shape of the entrance hole makes it very

Figure 154. The false exits of *Sarracenia minor* as viewed obliquely from below.

difficult for the imprisoned insect to escape. Eventually, it drowns in the digestive liquid contained within the leaf or dies of exhaustion. Since *S. psittacina* frequently grows in very wet habitat that is often inundated during heavy rainfall, the leaves are often submerged, one result of which is that they also trap small aquatic creatures.

The traps of *Sarracenia* most frequently catch ants, flies, wasps and bees, but they also trap beetles, slugs and snails. Even though *Sarracenia* populations frequently grow together in massive numbers and competition for insect prey is great, each leaf usually traps several insects, and often contain up to 35g (1.2 oz) of prey. Narcotic substances have been identified in the nectar of several species of *Sarracenia,* and these are believed to cause insects to become intoxicated and to more readily fall into the traps.

Figure 155. The unique traps of *Sarracenia psittacina* function both above and below water level.

Figure 156. This sectioned leaf of *Sarracenia psittacina* clearly shows the inward-projecting entrance at the hole of the trap.

Distribution

The genus *Sarracenia* is endemic to the United States and Canada (Figure 157). Most species in the genus occur within an arc from eastern Texas to Florida to New Jersey, and inland to Tennessee. The only taxon

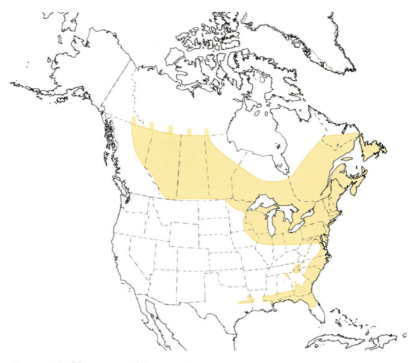

Figure 157. The range of *Sarracenia*.

that ranges beyond this area is *S. purpurea* ssp. *purpurea,* which occurs across the northeastern United States and most of the southern tier of provinces in Canada. There is overlap of the ranges of many species of *Sarracenia*, and hybrids frequently occur in these areas of sympatry.

Habitats

Sarracenia predominantly occur in marshy savannahs (Figure 158), open-canopy boggy pine forests, marl fens and alongside streams and lakes. Four characteristics are common to these various habitat types and are fundamental in understanding the growing requirements of *Sarracenia*.

1. *Sarracenia* consistently and exclusively grow where the substrate, or at least the subsurface substrate, is permanently moist or wet. *Sarracenia* habitat is always influenced by water from springs, seepage or small streams or it is located in low lying areas where runoff naturally accumulates. The

habitat of *Sarracenia* in the Gulf Coast and Atlantic Sea-board receives between 1200–2800 mm of precipitation annually. *Sarracenia* habitat does not require standing wa-ter, but populations are frequently observed growing in habi-tat that is temporarily flooded and submerged.

2. The healthiest *Sarracenia* populations with the most beau-tiful and brightly coloured leaves always occur in bright, sunny habitat. Since most *Sarracenia* occur between 30–40° North Latitude and grow predominantly in open habitat, the photoperiod is lengthy and lasts for up to fourteen hours each day during the height of summer. In 40% shaded con-ditions, the foliage of *Sarracenia* is usually dull green in colour and, in yet darker shade, the foliage is typically eti-olated and spindly.

3. The substrate of the *Sarracenia* habitat is almost always acidic (pH 4–5) and low in nitrates and phosphates. Usually it

Figure 158. The boggy, open habitat of *Sarracenia flava* and *Sarracenia leucophylla* in early spring.

consists of poorly drained sand and peaty humus that has accumulated in saturated conditions and been leached of nutrients.

4. The summer temperature across southern United States is warm-temperate. During June, July and August, the hottest months of the year, the daytime temperature peaks at between 25–30°C, although occasionally it will reach 35°C. *Sarracenia* vary in their tolerance of winter temperatures, but all species experience and survive frost and snowy conditions in at least part of their natural habitat. The more southerly species, *S. minor, S. alata, S. leucophylla, S. psittacina*, and *S. rubra*, do not do well in prolonged temperatures below -10°C. *S. purpurea* ssp *venosa, S. flava* and the montane *S. oreophila* are better adapted to cold and withstand temperatures as low as -15°C. *S. purpurea* ssp. *purpurea* is the most cold tolerant of all the *Sarracenia* and experiences prolonged icy conditions at least as low as -25°C in its northerly habitat.

General Ecology

Each species of *Sarracenia* has a slightly different annual growth cycle and most produce different types of leaves at different times over the course of the growing season. All species flower for 2–3 weeks during April and May. After flowering, *S. flava* and *S. oreophila* produce several tall and robust, primary carnivorous leaves, followed by a number of noncarnivorous leaves during the course of summer; these are ensiform in *S. flava* and sickle shaped in *S. oreophila*. Then in late summer and autumn these two species also produce smaller secondary carnivorous leaves before becoming dormant for the winter. *S. leucophylla* and *S. rubra* differ in that, after flowering, they first produce relatively small and spindly primary carnivorous leaves in spring and then produce ensiform noncarnivorous leaves over the summer, followed by large and robust secondary carnivorous leaves in late summer and autumn. *S. minor, S. psittacina* and *S. purpurea* produce pitchers of equal size throughout the growing season and do not produce any noncarnivorous leaves at all. *S. alata* produces carnivorous leaves in spring and autumn and a few short ensiform noncarnivorous leaves in summer. In this species the autumn leaves are only slightly smaller than the spring leaves. All *Sarracenia* species die down over winter and, in most, the foliage turns brown and dies back. The foliage of *S. psittacina* and *S. purpurea*, however, remains alive for 12–18 months.

A multitude of insects has evolved relations with *Sarracenia* species. Many moth caterpillars eat the leaves and flowers, and among the most voracious are the larvae of *Exyra semicrocea*. The caterpillars of this particular moth chew through the rigid veins and cause the leaves of the *Sarracenia* plants to bend over. Then, in the safety and protection of the bent leaf, the larvae eat the leaves from the inside out and pupate. Several species of praying mantises, spiders and frogs have learned that the pitchers are good hunting grounds and wait motionlessly on or just within the leaves to snatch visiting insects before they are trapped by the leaves (figures 159 and 160). Although these animals rob the plant of its meals, they do at least bring some benefits too; they typically are indiscriminate feeders and kill pests such as the adult *Exyra* moths.

Figure 159. A spider awaiting prey near the nectar bait of a *Sarracenia flava* leaf.

Figure 160. A praying mantis awaiting insects drawn to the nectar bait of a *Sarracenia leucophylla* leaf.

Species Descriptions

Sarracenia alata Wood

Original description: Wood, A., 1863, *Leaves and Flowers . . .*, 157.

The specific epithet *alata* is derived from the Latin *ala* (wing) and refers to the broad keel that extends down the front of the pitcher leaf of this species (Figure 161).

The range of *S. alata* is divided into two parts across the states of the Gulf Coast. The eastern part of the range encompasses extreme southwestern Alabama, southern Mississippi and a very small part of southeastern Louisiana. The western part of the range is located farther inland and extends from western Louisiana into eastern Texas (Figure 162).

Figure 161. A population of multicolored leaves of *Sarracenia alata* that are growing close together in Mississippi.

The flower is 4–7 cm wide. The petals are 2–3 cm long, rounded in shape and typically pale yellow or cream in colour (Figure 163). The sepals are slightly pointed and greenish yellow. The scape is 25–45 cm tall.

The pitchers of *S. alata* are extremely variable and many distinct colour varieties are discernable. No infraspecific taxa of *S. alata* have been formally classified or described. All infraspecific "variants" of *S. alata* are essentially the same as the "typical variant" in terms of morphology and size.

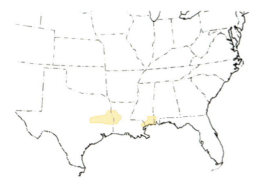

Figure 162. The range of *Sarracenia alata.*

Sarracenia alata 'Typical Variant'

The pitchers of the most widely distributed and typical variant of *S. alata* are predominately yellowish green and lined with a variable amount of red or purple venation which is concentrated around the pitcher opening (Figure 164). In some strains, light veins are also present on the interior of the tubular section of the leaf. Over the course of a summer the leaves darken in colour and suffuse orange or reddish as they age and die back.

Typically, the leaves are 20–60 cm tall and 3–5 cm wide at the pitcher opening. The very largest strains produce leaves in excess of 80 cm

Figure 163. The flower of *Sarracenia alata* 'Typical Variant.'

Figure 164. The leaf of *Sarracenia alata* 'Typical Variant.'

in height. The foliage is erect and tubular. The lower and middle sections of the leaf are narrow and slender, but it broadens in the upper section, especially towards the pitcher opening. The girth of the leaf is slightly constricted just below the pitcher opening, so that the upper parts of the leaf are slightly bulged in cross section. The lid is semicircular although slightly pointed towards the front of the leaf. It typically terminates in a 3–4-mm-long point. The ala (keel) is 2–12 mm broad and especially wide towards the base of the leaf. A very subtle pubescence that consists of 0.1–0.4-mm-long white hairs is often present on the exterior surface of the leaves. *S. alata* produces a few short ensiform, noncarnivorous leaves in summer.

In the western part of the species' range, some populations of *S. alata* consistently produce leaves that are about half the size of the typical variant, but are typical in all other respects. It is not clear if these populations represent ecophenes or a morphologically stable small variant of this species. In cultivation, this variant is usually called "stocky" or "short variety."

Sarracenia alata 'Areolated Variant' (Undescribed)

In this variant, subtle white areoles are present on the upper parts of the leaf, mainly on the lid and below the pitcher opening. The leaves are otherwise usually green or lined lightly with subtle red veins as in the typical variant. It is possible that the areoles are the result of ancestral hybridization with *S. leucophylla,* especially since in some cases the areolated specimens — or other *S. alata* plants in the same populations — display slightly red- or orange-tinted flowers. The range of *S. leucophylla* possibly once extended farther west through Mississippi and Louisiana than it does today, and the two species would have grown together and readily hybridized as they still do in Alabama. Hybrids in Alabama can be found that show exactly the same traits as the "areolated" *S. alata* from elsewhere in the species' range. This variant is sometimes called "areolata" although it has not yet been formally published or described under this name.

Sarracenia alata 'Copper Top Variant' (Undescribed)

In this variant, the upper side of the lid develops a vivid coppery-bronze colouration in a manner similar to *S. flava* var. *cuprea.* The remaining parts of the leaf are usually greenish yellow and lined with delicate purple veins.

Sarracenia alata 'Green Variant' (Undescribed)

The leaves of this variant appear pure yellowish green in colour. Subtle red colouration is present only on very young or developing leaves or on old pitchers that discolour as they die back in autumn (Figure 165). Recently, at least one distinct strain has been discovered that altogether

Figure 165. The leaf of *Sarracenia alata* 'Green Variant.'

Figure 166. The leaf of *Sarracenia alata* 'Purple Black Variant.'

lacks the ability to produce red colouration in its leaves. The pitchers of this strain are entirely yellowish green as in *S. purpurea* ssp. *purpurea* fm. *heterophylla*.

Sarracenia alata 'Pubescent Variant' (Undescribed)

In this variant, the exterior surface of the leaves are lined with a velvet coating of 0.2–2.2-mm-long silver hairs which give the leaves a velvet shine when viewed at an angle. It is not clear what function the hairs perform; perhaps they are a means of protection against insect predators or serve to dissipate the intensity of the sunlight, or possibly they are further evidence of ancestral hybridization with *S. leucophylla* as this characteristic is more strongly associated with that species. The pubescent trait is represented in all of the variants of *S. alata*. In particular, pubescent strains of the 'Purple Black Variant' are especially beautiful.

Sarracenia alata 'Purple Black Variant' (Undescribed)

The pitchers of this variant develop spectacular purplish-black colouration on the upper section of the pitcher, especially concentrated on the underside of the lid and around the pitcher opening (Figure 166).

Figure 167. The leaf of *Sarracenia alata* 'Red Lid Variant.'

Figure 168. The leaf of *Sarracenia alata* 'Red Variant.'

The colour of the pitchers is not truly black but usually a very dark shade of purple. This variant requires intense photoperiods in excess of ten hours per day in order to develop its distinctive colouration, otherwise the leaves only develop reddish-green colouration. In cultivation, root disturbance also affects the colour of the leaves and after a plant is repotted, it may take up to two years for colouration to stabilize. In cultivation this variant has informally been named "nigrapurpurea" and "black tube."

Sarracenia alata 'Red Lid Variant' (Undescribed)

In this variant, the underside of the lid suffuses vivid red in colour, often emphasised by prominent purple veins (Figure 167). The rest of the leaf typically remains greenish yellow lined with ornate but highly variable red veins. The upper surface of the lid suffuses deep copper or bronze.

Sarracenia alata 'Red Variant' (Undescribed)

The leaves of this variant develop striking uniform scarlet or deep-maroon colouration, usually lined with variable purple veins (Figure 168).

Figure 169. The leaves of *Sarracenia alata* 'Veined Variant.'

The colour of the pitchers darkens over the course of summer and turns purplish by autumn. A variant that retains a yellow lid, as in *S. flava* var. *rubricorpora*, does not exist in *S. alata*.

Sarracenia alata 'Veined Variant' (Undescribed)

In this variant, the exterior surface of the leaves is netted with intricate red or purple veins (Figure 169). The remaining parts of the leaves are usually greenish yellow, and often the underside of the lid suffuses pure red. Strains with particularly intense venation appear almost

Figure 170. The flower of *Sarracenia alata* 'White Flowered Variant.'

solid red. In rare cases, some strains of *S. alata* develop intense red vein-ing only on the interior surface of the leaves. The red interior contrasts with the pure green exterior of the leaf and is especially beautiful.

Sarracenia alata 'White Flowered Variant' (Undescribed)

A small proportion of all known variants of *S. alata* produce petals that are extremely light cream or pure white in colour rather than the typical pale yellow (Figure 170). This unusual white floral colouration is a stable genetic trait and is readily inherited in offspring. In localized or isolated populations, the white flower trait can be prevalent, although in terms of its presence throughout the total population of this species, it is relatively rare. Hybrids involving this variant and red-flowered species of *Sarracenia* have not yet been produced in cultivation but, if and when done, might yield an interesting floral colouration.

Sarracenia flava Linnaeus

Original description: Linnaeus, C., 1753, *Species Plantarum* 1: 510.

The specific epithet *flava* is derived from the Latin *flava* (yellow) and refers to the colouration of the leaves and flowers of this species.

The range of *S. flava* extends south in an arc from southeastern Virginia across the Atlantic and Gulf coastal plains of North Carolina, South Carolina, Georgia, western Florida and southeastern Alabama (Figure 171). An isolated inland population also exists on the Piedmont of northwestern North Carolina.

The flower of *S. flava* is 4–7 cm in diameter, the petals are strap shaped with rounded ends, 3–5 cm in length and bright yellow in colour (Figure 172). The sepals and the umbrella-shaped pistil are yellowish green. In some colour varieties the petals develop light red veins or flecks. The flower of *S. flava* is one of the largest and most spectacular in the genus.

The pitchers of *S. flava* (Figure 173) are the most variable of all the *Sarracenia* species and a wide spectrum of beautiful colour varieties has been described and named. Throughout all of the colour varieties, the structure and size of the foliage is roughly the same as in *S. flava* var. *flava*.

Sarracenia flava var. *flava*

The leaves of *S. flava* var. *flava* are predominantly yellowish green in colour but have a small reddish-purple blotch present at the back of the pitcher opening (Figure 174). Variable purple or red veins radiate across the underside of the lid. *S. flava* var. *flava* occur predominantly in the northern part of the species' range, especially in North Carolina and Virginia where it is the most prevalent variety.

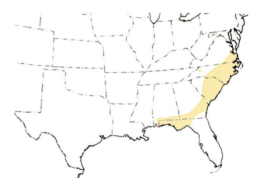

Figure 171. The range of *Sarracenia flava*.

Figure 172. The flower of *Sarracenia flava*.

The pitchers of the typical variety are 40–70 cm in length, although the largest strains produce pitchers in excess of 100 cm in height. The foliage is erect and tubular. The lower and middle sections of the leaf are slender. The upper parts of the leaf are infundibular, especially towards the pitcher opening. The lid is circular and broad. The ala is 2–8 mm wide, narrow in the upper parts of the leaf and broad at the base. *S. flava* produces several large ensiform noncarnivorous leaves during the course of summer.

Sarracenia flava var. *atropurpurea* (Bull) Bell

> **Original description:** Bell, C., 1949, *Journal of the Elisha Mitchell Scientific Society* 65: 137-166.

The varietal epithet *atropurpurea* is derived from the Latin *ater* (dark) and *purpureus* (purple coloured) and refers to the uniform red-to-purple colouration characteristic of the leaves of this variety (Figure 175).

Figure 174. The leaves of *Sarracenia flava* var. *flava*.

Figure 175. The leaf of *Sarracenia flava* var. *atropurpurea*.

Variable dark-purple veins dominate the interior of the pitcher opening and the underside of the lid. This variety is distinguished from *S. flava* var. *rubricorpora* by the pure red colouration of the lid. *S. flava* var. *atropurpurea* requires intense photoperiods in excess of ten hours per day in order to develop its pure red colouration, otherwise the leaves are coloured reddish green. The noncarnivorous leaves of this variety often are also reddish. This variety occurs throughout the range of *S. flava*, but it is particularily prevalent in localized populations in Florida.

Sarracenia flava var. *cuprea* Schnell

Original description: Schnell, D. E., 1998, *Carnivorous Plant Newsletter* 27: 116–120.

The varietal epithet *cuprea* is derived from the Latin *cuprea* (copper coloured) and refers to the bronzy copper colouration of the upper

Figure 173 (facing page). The majestic leaves of *Sarracenia flava* var. *rugelii* growing in the Florida Panhandle. This strain is also the cultivar *Sarracenia maxima*.

Figure 176. The leaf of *Sarracenia flava* var. *cuprea*, viewing the upper surface of the lid.

Figure 177. The leaf of *Sarracenia flava* var. *maxima*.

surface of the lid (Figure 176). The remaining parts of the leaf are predominantly yellowish green and lined with delicate purple veins to varying degrees, especially on the interior of the pitcher opening and the underside of the lid. In some specimens, the copper colouration of the lid darkens to an attractive red colour. *S. flava* var. *cuprea* occurs predominantly in Virginia, North Carolina, South Carolina and Georgia. It is extremely rare or possibly absent in the Gulf Coast part of the species' range.

Sarracenia flava var. *maxima* Bull ex Masters

Original description: Masters, M. T., 1881a, *Gardeners' Chronicle, 2nd Series* 15: 817–818.

The varietal epithet *maxima* is derived from the Latin *maximus* (large or largest), but its application to this taxon is misleading since the leaves of this variety are not actually larger than, or structurally different from, any of the other varieties of *S. flava* (Figure 177). The truly distinguishing characteristic of this variety is the pure yellowish-green colouration of the foliage. The leaves are not altogether free of red colouration; the young developing pitchers are tinted slightly orange and the aging leaves suffuse reddish as they die back in autumn. Unlike most other *Sarracenia* species, a strain of *S. flava* that is entirely devoid of red pigmentation has not yet been discovered. *S. flava* var. *maxima* occurs mainly in North Carolina, South Carolina, Virginia and Georgia. It is extremely rare or possibly absent in the Gulf Coast part of the species' range.

Sarracenia flava var. *ornata* Bull ex Masters

Original description: Masters, M. T., 1881a, *Gardeners' Chronicle, 2nd Series,* 15: 628–629.

The varietal epithet *ornata* is derived from the Latin *ornate* (ornately, splendidly) and refers to the beautiful colouration of the pitchers in this taxon. The leaves are yellowish green and lined with vibrant red or purple veins, usually concentrated around the pitcher opening and on the lid (Figure 178). In some strains, the interior of the pitcher opening suffuses pure orange, red or purple in colour. The veins vary greatly in pattern and degree. *S. flava* var. *ornata* occurs throughout the species' range.

Figure 178. The leaf of *Sarracenia flava* var. *ornata*.

Sarracenia flava var. *rubricorpora* Schnell

Original description: Schnell, D. E., 1998, *Carnivorous Plant Newsletter* 27: 116–120.

The varietal epithet *rubricorpora* is derived from the Latin *rubra* (red) and *corpus* (body) and refers to the colouration of the leaf. The tubular body of the pitcher is pure red in colour, while the lid remains predominantly yellow, lined with variable red veins (figures 179 and 180). A variable dark-purple blotch. similar to that in *S. flava* var. *rugelii*, is present at the back of the pitcher opening from which variable purple veins radiate. In some strains the lower parts of the leaf have a bluish-purple blush. The colouration of *S. flava* var. *rubricorpora* is extremely variable. In some strains the lid lacks red veins altogether and is pure yellow while at the other end of the spectrum, the lid and pitcher opening can be so intricately veined that they appear almost pure red. The colouration of the tubular portion of the leaf varies in a wide spectrum

Figure 179. The leaves of *Sarracenia flava* var. *rubricorpora* exhibit a wide range of colouration. The specimen in the centre displays typical colouration for this variety, whereas the other specimens represent extremes of the continuous range of variation observed in this variety.

Figure 180. The leaf of *Sarracenia flava* var. *rubricorpora*.

of shades from coppery orange, through scarlet and crimson, to intense dark purple. The true purple strains are extremely rare and originally occurred at just a few sites in the Florida Panhandle — most of which have now been destroyed by urban development. The colour of the purple variant of *S. flava* var. *rubricorpora* is distinct and this variant might deserve formal classification as an independent variety.

S. flava var. *rubricorpora* develops its excellent red colouration only when it occurs in open habitat exposed to full sunlight. In shaded conditions the leaves only turn reddish green. As in all colour forms of *Sarracenia* species, root disturbance also affects the colour of the leaves and repotted plants can take up to two years for colouration to stabilize. *S. flava* var. *rubricorpora* has a tendency to grow individually rather than to form dense clumps. In bright conditions, the noncarnivorous leaves of this variety suffuse reddish. *S. flava* var. *rubricorpora* is distributed mainly in the south of the species' territory, predominantly across northern Florida.

Figure 181. The leaf of *Sarracenia flava* var. *rugelii*.

Sarracenia flava var. *rugelii* (Shuttleworth ex Alphonse de Candolle) Masters

Original description: Masters, M. T., 1881b, *Gardeners' Chronicle, 2nd Series*, 16: 11–12, 40–41.

The varietal epithet *rugelii* honours the 19th-Century botanist Ferdinand Rugel. In this variety the pitchers are pure yellowish green in colour apart from a prominent reddish-purple blotch present on the back of the pitcher opening (figures 181 and 182). The leaves open greenish in colour but turn vibrant golden yellow if exposed to direct sunlight. The purple colouration is quite variable; in some strains it dominates the entire back of the pitcher opening, but in others it is greatly reduced. The red colouration of this variety is readily inherited in hybrids, especially *S. flava x leucophylla*. *S. flava* var. *rugelii* is among the most vigorous varieties and the leaves are often in excess of 80 cm in length. More than other varieties of *S. flava*, this one has a tendency to form clumps up to 1 m across. It is extremely common in the southwestern part of the species' range, especially through the Florida Panhandle and Alabama, but it is relatively rare in Georgia, South Carolina, North Carolina and Virginia. Intergrades with the other varieties, especially *S. flava* var. *maxima*, are common.

Sarracenia leucophylla Rafinesque

Original description: Rafinesque, C. S., 1817, *Florula Ludoviciana*, 14.

The specific epithet *leucophylla* is derived from the Greek *leukos* (white, pale) and *phyllon* (leaf) and refers to the colouration of the foliage (Figure 183).

The range of *S. leucophylla* extends from the extreme southwestern corner of Georgia across the Florida Panhandle and southern Alabama to the southeastern corner of Mississippi (Figure 184).

The flower of *S. leucophylla* is 4–7 cm in diameter. The petals are strap shaped with rounded ends, 3–5 cm in length and bright crimson to maroon in colour (Figure 185). The sepals are pointed and also maroon in colour. The umbrella-shaped pistil is reddish green. Unique within the genus, the seed pods of *S. leucophylla* split from the rear.

The pitchers of *S. leucophylla* are extremely variable and a number of distinct colour varieties can be distinguished. None of the infraspecific taxa of *S. leucophylla* have been formally classified or described, partly because the colour varieties of *S. leucophylla* mostly form a continuous spectrum of diversity which makes this species more difficult than any other of *Sarracenia* to subdivide infraspecifically. All infraspecific variants of *S. leucophylla* are essentially the same as the typical variant in terms of morphology and size.

Figure 182 (following pages). A population of *Sarracenia flava* var. *rugelii* flowering in spring in northern Florida.

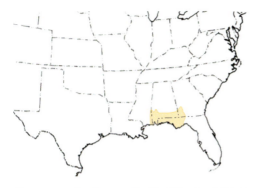

Figure 184. The range of *Sarracenia leucophylla.*

Sarracenia leucophylla 'Typical Variant'

In the most widely distributed and common variant of *S. leucophylla*, the uppermost parts of the leaves are lined with variable white areolation and netted with conspicuous red and green veins (Figure 186). The remaining parts of the leaves are predominantly reddish green. The foliage is erect and tubular, and each leaf is typically 40–70 cm tall, although the largest strains produce leaves in excess of 100 cm tall. The lower and middle sections of the leaf are narrow and slender and the upper parts are infundibular, especially towards the pitcher opening. The lid is broad and undulates greatly. The ala is 2–8 mm wide and narrow in the upper parts of the pitcher but broad towards its base. *S. leucophylla* produces several large ensiform noncarnivorous leaves during early summer.

Sarracenia leucophylla 'Green and White Variant' (Undescribed)

In this variant, the leaves lack visible red colouration even when the plant grows in direct sunlight (Figure 187). The white areoles are netted with pure green veins. Red colouration is visible only on very young, developing leaves, on aging leaves that are dying back and on the flowers which are pure red in colour and identical to the typical variant of the species. At least three distinct red-pigment-free strains of *S. leucophylla* have been discovered that are altogether free of red colouration and produce leaves that are pure green and white and flowers that are yellowish green.

Figure 183 (facing page). *Sarracenia leucophylla* plants growing amidst wildflowers in southern Alabama.

Figure 185. The flower of *Sarracenia leucophylla*.

Figure 186. The leaf of *Sarracenia leucophylla* 'Typical Variant.'

Figure 187. The leaf of *Sarracenia leucophylla* 'Green and White Variant.'

Figure 188. The leaf of *Sarracenia leucophylla* 'Purple and White Variant.'

Figure 189. The leaf of *Sarracenia leucophylla* 'Red Tube Variant.'

Sarracenia leucophylla 'Pubescent Variant' (Undescribed)

In this variant, the exterior surface of the foliage is lined with a velvety coating of 0.3–2.5-mm-long, soft, white hairs that give the leaves a silvery sheen when viewed from an angle. This variant of *S. leucophylla* has the longest, most prominent hairs of all of pubescent varieties of the various *Sarracenia* species.

Sarracenia leucophylla 'Purple and White Variant' (Undescribed)

In this variant, the veining on the upper section of the leaves is a dark shade of purplish black, which contrasts beautifully with the pure white areolation (Figure 188). The lower parts of the leaf are typically pure yellowish green or lightly lined with dark veins.

Sarracenia leucophylla 'Red Tube Variant' (Undescribed)

In this variant, almost all parts of the leaf develop striking bright-red colouration (Figure 189). The veins are vivid scarlet and the areolation suffuses a vibrant shade of pink. In some strains the areoles on the upper surface of the lid remain pure white and contrast beautifully with the rest of the leaf. The very base of the pitcher usually remains greenish. The leaves of this variant tend to be tall and slender. This variant is extremely rare in the wild and the most colourful strains in cultivation generally result from selective breeding programmes.

Sarracenia leucophylla 'Red and White Variant' (Undescribed)

In this variant, the veining on the upper section of the leaf is bright red in contrast to the areoles which remain pure white (Figure 190). The lower parts of the leaf are reddish green as in the typical variant. In some strains, the rim of the pitcher opening, the "lip," suffuses bright pink.

Sarracenia leucophylla 'White Variant' (Undescribed)

The leaves of this variant develop extremely faint veins so that the white areolation is particularly prominent (Figure 191). In the most extreme strains, the upper section of the pitcher can appear virtually pure white. The lower parts of the leaves are greenish in colour similar to the typical variant. It is relatively rare overall. In cultivation this variant has been informally named "alba," which is derived from the Latin *albus* (white).

Sarracenia leucophylla 'Yellow Flowered Variant' (Undescribed)

Several strains of *S. leucophylla* have been discovered that produce flowers that are pure yellowish green in colour. The leaves of the different strains usually have minimal red colouration, although they are not devoid of red pigment. The most widely grown strain of this variant is the cultivar *S.* 'Schnell's Ghost.'

Figure 190. Three expressions of the leaves of *Sarracenia leucophylla* 'Red and White Variant.'

Sarracenia minor Walter

Original description: Walter, T., 1788, *Flora Caroliniana, secundum*, 153.

The specific epithet *minor* is derived from the Latin *minor* (smaller) and refers to the typical short size of the leaves in this species.

The range of *S. minor* extends southward across the Atlantic and Gulf coastal plains from southern North Carolina across South Carolina, Georgia and northern Florida (Figure 192).

The flower of *S. minor* is 3–6 cm in diameter. The slightly pointed sepals and the 2–3-cm-long, rounded petals are pale yellow in colour (Figure 193). The umbrella-shaped pistil is yellowish green.

The pitchers of *S. minor* vary subtly in terms of colour and size. Two distinct varieties of *S. minor* have been identified and described and one further "variant" is discernable although it remains unclassified.

Sarracenia minor var. minor

The leaves of the typical variety of *S. minor* are 15–25 cm tall and 1.5–3 cm wide at the pitcher opening (figures 194 and 195). The foliage is erect and tubular. Unlike all other *Sarracenia* species, the lid curls forward over the pitcher opening similar to the cowled hood of a monk's cloak. The lower and middle sections of the leaf are narrow and slender, the upper section of the leaf broadens greatly towards the upper parts of the leaf. The ala is 2–20 mm wide, it is narrow below the pitcher opening but broadens towards the base of the leaf. The flowers of *S. minor* var. *minor*

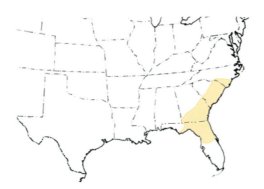

Figure 192. The range of *Sarracenia minor.*

Figure 191. The particularly striking leaf of *Sarracenia leucophylla* 'White Variant.'

Figure 193. The flower of *Sarracenia minor* var. *minor.*

open on a scape that is roughly as tall as the pitchers. No noncarnivorous leaves are produced by *S. minor*. The leaves of *S. minor* var. *minor* are predominantly yellowish green apart from the hood-shaped lid which suffuses coppery pink to dark red. The rim, or lips, of the pitcher opening is typically pure red and the back section of the upper third of the leaf is mosaicked with variable white areolation.

Sarracenia minor var. *okefenokeensis* Schnell

Original description: Schnell, D. E., 2002b, *Carnivorous Plant Newsletter* 31(2): 36–39.

The varietal epithet *okefenokeensis* refers to the Okefenokee Swamp in Georgia where this giant variety predominantly occurs. *S. minor* var. *okefenokeensis* differs from *S. minor* var. *minor* in that the pitchers are commonly in excess of 100 cm in height. The tubular section of the leaf

Figure 194 (facing page). The flowers and leaves of *Sarracenia minor* var. *minor.*

Figure 195. The leaves of *Sarracenia minor.*

is greatly elongated and the overall size of the pitcher opening is en-larged. The foliage of this variety is commonly much more colourful than that of the typical variety and in some strains the leaves suffuse pure reddish orange with pink-tinted areoles. The flowers of this variety are born on relatively elongated scapes, usually 25–75 cm tall.

Sarracenia minor 'Green and White Variant' (Undescribed)

Multiple strains of *S. minor* have been discovered that are entirely free of red colouration and consequently the flowers and foliage appear pure yellowish green in colour, with the exception of the white areoles. Other than the colouration, this variant is identical to the typical variety.

Sarracenia oreophila (Kearney) Wherry

Original description: Wherry, E., 1933, *Bartonia* 15: 1–8.

The specific epithet *oreophila* is derived from the Greek *oros* (mountain) and *filos* (friend) and refers to the presence of this species in montane habitat.

The range of *S. oreophila* consists of a small number of disjunct units scattered across the Appalachian Highlands and adjoining lowlands. The largest of these units occurs in the southern Ridge and Valley physiographic province of northeastern Alabama, with others being located in central Alabama, southwestern Georgia, astride the Georgia-North Carolina border and in northeastern Tennessee (Figure 196). Across its small range, *S. oreophila* is highly endangered and is included both in *CITES Appendix I* and the *List of Endangered and Threatened Plants* of the United States.

The flower of *S. oreophila* (Figure 197) is 3–6 cm in diameter and is similar to that of *S. flava*. The slightly pointed sepals and the 3–4-cm-long, strap-shaped petals are bright yellow in colour. The umbrella-shaped pistil is yellowish green. The scape is typically 20–30 cm tall.

Despite the small and highly fragmented range of *S. oreophila*, a relatively large degree of diversity was probably originally discernible as in all other species of *Sarracenia*. Unfortunately, the vast majority of wild populations of this species were destroyed before any infraspecific taxa were recognized or described. As a consequence, it is difficult to appreciate what variants of this species might originally have existed, and so here we can consider the diversity of *S. oreophila* only in very general terms.

The pitchers of *S. oreophila* are typically 25–40 cm tall and 2–4 cm wide at the pitcher opening. The foliage is erect and tubular, although

Figure 196. The range of *Sarracenia oreophila*.

Figure 197. The flower of *Sarracenia oreophila*.

often slightly crescent shaped in cross section (figures 198 and 199). The lower and middle sections of the leaf are narrow, but it broadens towards its upper parts. The lid is roughly circular in shape. The back of the pitcher opening is wider than in *S. flava*. The ala is 2–10 mm wide; it is narrow in the upper section but broadens towards the base of the leaf.

   S. oreophila has evolved a unique annual growth cycle to suit the unusual conditions of its habitat. The availability of water at the predominently upland areas where this species naturally occurs differs profoundly over the course of a year. Precipitation is concentrated during spring and summer, although excessive drawdown, evaporation and transpiration during June, July and August causes a pronounced water deficit during the summer months. In response to the dry conditions of summer, *S. oreophila* first produces a succession of large primary carnivorous leaves after flowering during April and May, but these die back suddenly over the course of July and August and, in their place, several small, inconspicuous sickle-shaped, noncarnivorous leaves emerge. Several weeks later, when conditions have improved during autumn, smaller secondary carnivorous leaves

Figure 198 (facing page). The leaves of *Sarracenia oreophila*.

Figure 199. The leaves of *Sarracenia oreophila* (left) showing typical colouration and (right) exhibiting pronounced venation.

develop and catch prey for a few weeks during September and early October before winter and the onset of dormancy. In cultivation, *S. oreophila* follows this pattern of growth regardless of the availability of water. The primary leaves of *S. oreophila* are typically yellowish green and lined with delicate purplish-black veins in the pitcher opening. Often the veins are most prominent on the interior surface of the leaf and in some specimens the venation is greatly exaggerated so that the leaves are lined with conspicuous purple veins similar to those of *S. flava* var. *ornata*. In other strains, the interior surface of the pitcher opening suffuses bright red or purple.

Sarracenia psittacina Michaux

Original description: Michaux, A., 1803, *Flora Boreali-Americana* 1: 311.

The specific epithet *psittacina* is derived from the Greek *psittakos* (parrot) and refers to the shape of the leaf which is reminiscent of the beak of a parrot.

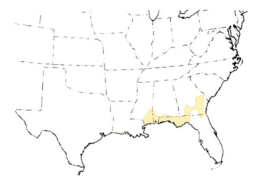

Figure 200. The range of *Sarracenia psittacina.*

The range of *S. psittacina* extends from southeastern Georgia southwest to the Florida Panhandle, then west to extreme southeastern Louisiana (Figure 200). *S. psittacina* often grows at the margins of ponds or small streams and frequently grows semi-aquatically when its habitat is temporarily flooded.

The flower of *S. psittacina* is 2–4 cm in diameter and borne on a tall, 25–45-cm-long scape (Figure 201). The short 2–3-cm-long petals are dark red or purple and rounded in shape. The sepals are the same

Figure 201. The flower of *Sarracenia psittacina.*

colour and slightly pointed. The umbrella-shaped pistil is reddish green. In proportion to the length of the foliage, the scape of *S. psittacina* is the largest in the genus. Since the foliage of this species is readily concealed by low growing vegetation, the scapes of *S. psittacina* are frequently the only visible indication of the presence of this species within its habitat.

The pitchers of *S. psittacina* are the most distinctively shaped within the genus — and in many respects parallel those of *D. californica* — especially in terms of morphology and trapping processes. It remains unclear whether the similarities between the two species represent parallel evolution or provide direct evidence of the common ancestery of the two genera. Significant differences between the two genera can, however, be identified — these include the structure of the transparent fenestration of *D. californica* and the white areolation of *S. psittacina* as well as the processes through which the foliage of the two species form; the leaves of *D. californica* twist as they enlarge and inflate whereas those of *S. psittacina* emerge and expand straight. None of the infraspecific variants of *S. psittacina* have been formally classified or described, but at least two variants are discernable in addition to the type form.

Sarracenia psittacina 'Typical Variant'

The leaves of typical *S. psittacina* consist of a tubular shaft that terminates in a hollow spherical structure (figures 202 and 203). The only entrance to the trap is a small inward-protruding hole on the centreward side of the inflated, dome-shaped end. The foliage is arranged in a compact horizontal rosette and rests directly on the surface of the ground. Each pitcher is 10–25 cm in length and 1–4 cm wide. The ala is 2–20 mm wide, narrow at the base of the leaf and broad towards the dome. The foliage is predominantly reddish green and lined with red veins and variable white areoles. Along the tubular shaft of the leaf, the areolation is elongated. *S. psittacina* does not produce any noncarnivorous leaves. It is highly tolerant of shade and grows healthily in 0–50% shaded conditions, but in 20–40% shade the foliage is predominantly green.

Figure 202 (facing page). *Sarracenia psittacina* growing in northern Florida.

Figure 203. The typical growing habit of *Sarracenia psittacina.*

Sarracenia psittacina 'Green and White Variant' (Undescribed)

Many strains of *S. psittacina* have been discovered that are entirely free of red colouration, and consequently the flowers and foliage appear pure yellowish green in colour, excepting the white areoles. Other than the colouration, this variant is identical to the typical variety.

Sarracenia psittacina 'Yellow Flowered Variant' (Undescribed)

Various strains of *S. psittacina* have been discovered which produce pure yellowish-green or yellowish-orange flowers, but in all other respects are typical. The floral colouration is attributed to a faulty gene, but this gene appears to be stable and can be inherited if two yellow-flowered strains are cross-pollinated.

Sarracenia purpurea Linnaeus

Original description: Linnaeus, C., 1753, *Species Plantarum* 1: 510.

The specific epithet *purpurea* is derived from the Latin *purpureus* (purple), and refers to the typical colouration of the flower and foliage of this species.

The range of *S. purpurea* encompasses the largest area of all of the American pitcher plants. At the northern edge of its range, *S. purpurea* occurs across a vast swath of the southern tier of Canadian provinces, from northeastern British Colombia through most of the Priarie Provinces and Ontario, southern Quebec, southern Newfoundland (including the island of Newfoundland and southern Labrador) and all of the Maritimes (Figure 204). This vast expanse continues southward into the United States to incorporate all of the New England states, most or all of the Great Lakes states, and all or part of Iowa, West Virginia, Maryland,

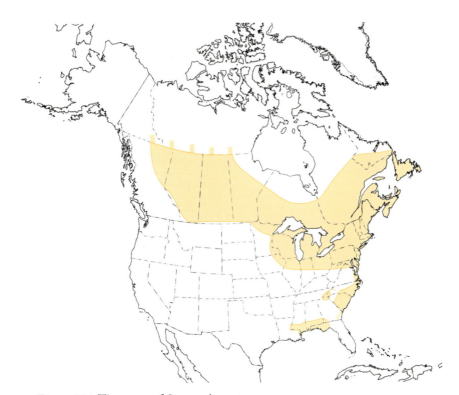

Figure 204. The range of *Sarracenia purpurea*.

New Jersey and Delaware. In the southern Middle Atlantic region, the range becomes markedly diminished. With minor exceptions, from Virginia south the range is decidedly associated with the Atlantic and eastern Gulf coastal plains. Disjunct populations occur in the interior of western North Carolina, including the higher elevations of the southern Appalachian Mountains and the immediately adjoining areas of adjacent states.

The northern part of the range, including all of the Canadian portion and that part in the United States south to the northern Middle Atlantic region, is occupied by the northerly subspecies, *S. purpurea* ssp. *purpurea*. The species' range from Virginia south, including the interior parts, is occupied solely by the southern subspecies, *S. purpurea* ssp. *venosa*. The two subspecies are sympatric in New Jersey and Delaware.

S. purpurea ssp. *purpurea* is very much adapted to the cold northern winters and endures temperatures at least as low as -25°C. The more southerly *S. purpurea* ssp. *venosa* is not quite as cold tolerant but nonetheless endures temperatures at least as cold as -10°C. Both subspecies are relatively tolerant of shaded conditions and readily grow in 0–50% shaded conditions, although in 20–40% shade the leaves of all varieties of the species are predominantly green in colour.

The flower is virtually the same in the two subspecies although generally it is proportionately larger in *S. purpurea* ssp. *venosa* var. *burkii*. The bloom is typically 3–5 cm in diameter and borne on a 20–40-cm-tall scape (Figure 205). The petals are short, 2–3 cm long, rounded in shape and reddish purple in colour. The sepals are pointed and purplish. The umbrella-shaped pistil is typically reddish green.

The pitchers of *S. purpurea* vary profoundly between the two subspecies, but the defining characteristics of each are remarkably constant across the species' vast range. *S. purpurea* is the most extensively studied of all *Sarracenia* and all known valid infraspecific taxa have been described and named.

Sarracenia purpurea ssp. *purpurea*

The pitchers of *S. purpurea* ssp. *purpurea* are 10–15 cm long and 2–4 cm wide at the pitcher opening (Figure 206). The lower and middle sections of the leaf are infundibular, but the girth is constricted just below the pitcher opening so that the tubular portion of the leaf appears slightly ventricose in shape. The lid is upright and affords little protection from

Figure 205. The flower of *Sarracenia purpurea*.

rain. The perimeter of the lid undulates to varying degrees and the interior surface is lined with 2–5-mm-long, downwards-pointing hairs. The lid is roughly semi-circular and curls around the sides of the pitcher opening so that it is somewhat conical in shape. The ala is 2–12 mm broad, narrow beneath the pitcher opening and at the base of the leaf and widest at the midsection. The exterior and interior of the leaves of *S. purpurea* ssp. *purpurea* are lined with a protective waxy cuticle and, consequently, the exterior of the leaves appear unusually smooth and shiny in appearance. The leaves of this subspecies are rigid and stiff and support their own weight even when completely full with digestive liquid. The leaves of *S. purpurea* ssp. *purpurea* are arranged in a compact circular rosette. In cross section, the leaf is crescent shaped which enables the lower portions of the leaves to rest directly on the ground. Each rosette consists of 5–20 leaves. Individual plants enlarge by dividing, and over several years, or possibly decades, one plant might eventually form a dense clump up to 120 cm across consisting of up to 150 individual growth points. The colouration of the foliage varies extensively among differing strains. Generally, however, the leaves of *S. purpurea* ssp. *purpurea* are predominantly reddish purple in colour. The lid and ala are typically yellowish green and lined with variable red veins.

Figure 206. The leaves of *Sarracenia purpurea* ssp. *purpurea*.

Sarracenia purpurea ssp. *purpurea* fm. *heterophylla*
(Eaton) Fernald

Original description: Fernald, M., 1922, *Rhodora* 24: 165–183.

The formal epithet *heterophylla* is derived from the Greek *heteros* (different) and *phyllon* (leaf) and refers to the unusual colouration of the foliage.

The pitchers of this form differ from those of *S. purpurea* ssp. *purpurea* in the respect that they altogether lack red pigmentation and all parts of the foliage and flowers are pure yellowish green in colour (Figure 207). Other than colouration, all aspects of this form are identical to the typical variety.

S. purpurea ssp. *purpurea* fm. *heterophylla* is rare in wild populations overall, but locally it can occur frequently or even dominantly. The yellowish-green colouration results from the lack of a gene that controls the red pigmentation, a condition that is consistently passed down in offspring if two *S. purpurea* ssp. *purpurea* fm. *heterophylla* plants cross-pollinate. This green-colouration trait is recessive, so if a *S. purpurea* ssp. *purpurea* fm. *heterophylla* plant cross pollinates with a pure *S. purpurea* ssp. *purpurea* with typical red-coloured leaves, the resultant offspring

Figure 207. The leaves of *Sarracenia purpurea* ssp. *purpurea* fm. *heterophylla.*

will consistently bare red-coloured foliage. However, the green colouration will be passively carried so that if the red coloured offspring were cross pollinated or self-pollinated, a proportion of the next generation would be pure green and be *S. purpurea* ssp. *purpurea* fm. *heterophylla.* In this way *S. purpurea* ssp. *purpurea* fm. *heterophylla* can spontaneously, and seemingly randomly, occur in populations of typical coloured red carrier plants, even if no pure *S. purpurea* ssp. *purpurea* fm. *heterophylla* plants have grown in the population for generations.

Sarracenia purpurea ssp. *purpurea* 'Veinless Variant'

(Undescribed)

Hybrids between *S. purpurea* ssp. *purpurea* and *S. purpurea* ssp. *purpurea* fm. *heterophylla* occasionally occur in the wild where populations of the two taxa grow in close proximity. The crossbreed offspring is distinguished by unique colouration which is predominantly yellowish green, but also tinged reddish orange, yet lacks the red venation that is typical of *S. purpurea* ssp. *purpurea* (Figure 208). Although not technically an infraspecific taxon, this unusual crossbreed has been known to stabilize and proliferate in localized, isolated populations and is informally called the 'Veinless Form.'

Figure 208. The leaves of *Sarracenia purpurea* ssp. *purpurea* 'Veinless Variant.'

Sarracenia purpurea ssp. *venosa* (Rafinesque) Wherry

Original description: Wherry, E., 1933, *Bartonia* 15: 1–8.

The subspecific epithet *venosa* is derived from the Latin *vena* (vein) and refers to the prominent venation typical of the leaves of this taxon. The pitchers are 10–25 cm long, 3–6 cm wide and proportionately stout and bulbous (Figure 209). The lower and middle sections of the leaves are infundibular, but the girth is constricted just below the pitcher opening so that the tubular portion of the leaf appears ventricose in shape. The lid is proportionately large and conspicuous, and in some strains undulates greatly. The interior surface of the lid is lined with 3–6-mm-long, downwards-pointing hairs. The ala is 2–16 mm broad, but narrow beneath the pitcher opening and at the base of the leaf and widest at the midsection. The exterior of the leaf is lined with 0.2–1-mm-long hairs which give the pitchers a soft velvety feel. The leaves of *S. purpurea* ssp. *venosa* are less rigid than those of *S. purpurea* ssp. *purpurea*. The leaves rest directly on the ground and are unable to support their own weight. The foliage is arranged in a circular rosette and each plant consists of

Figure 209 (facing page). The leaves of *Sarracenia purpurea* ssp. *venosa*.

2–10 (in the field usually 2–6) leaves. *S. purpurea* ssp. *venosa* does not form large clumps, but grows individually or in small groups.

In sunny habitat, 0–20% shade, the leaves are generally reddish purple and lined with dark red veins, especially on the interior of the lid. In 20–45% shade, however, the leaves are predominantly yellowish green and netted with purple veins, and the length of the leaf and the size of the ala is increased. In yet denser shade, the leaf is etiolated and the tubular aspect of the foliage is lost. The most colourful and vigorous specimens always grow in direct sunlight.

Sarracenia purpurea ssp. *venosa* var. *burkii* (Schnell)
Original description: Schnell, D. E., 1993, *Rhodora* 95: 6–10.

The varietal epithet *burkii* honours the American horticulturist Louis Burke, who cultivated this variety for the first time during the 1930s. *S. purpurea* ssp. *venosa* var. *burkii* is distinguished by its beautiful pink or pastel-pink flowers (Figure 210), rather than the usual red or dark purple floral colouration. In some strains, the petals are such a light shade of pink that they appear almost pure white in colour. The umbrella-shaped pistil is consistently pastel yellowish green in colour (Figure 211).

The flower of *S. purpurea* ssp. *venosa* var. *burkii* is 5–8 cm in diameter and generally much larger than the bloom of the typical variety of *S. purpurea* ssp. *venosa*. The pitcher also differs slightly in appearance and is usually more bulbous and proportionately larger in this variety than is typical of other taxa in the subspecies (Figure 212). Also in many strains of *S. purpurea* ssp. *venosa* var. *burkii*, the lid undulates more dramatically and can curl around the sides of the pitcher opening to a greater degree than is seen in *S. purpurea* ssp. *venosa*. The colouration of the leaf in this variety is more variable than in all other varieties of *S. purpurea* ssp. *venosa* and some strains exhibit pure purple, red, pinkish or even orangey colouration. Schnell (2002a) suggested that this variety may be the only representation of *S. purpurea* ssp. *venosa* in the Gulf Coast part of the species' range, and this appears to be true. It also is apparent that this variety does not occur in any other part of this subspecies' range.

Figure 210. The flower of *Sarracenia purpurea* ssp. *venosa* var. *burkii* on plants growing in northern Florida.

Figure 211. The flower of *Sarracenia purpurea* ssp. *venosa* var. *burkii* showing (left) the umbrella-shaped pistil and (right) the pale petals that are typical of the variety.

Sarracenia purpurea ssp. *venosa* var. *burkii* fm. *luteola* Hanrahan and Miller

Original description: Hanrahan, R., and J. Miller, 1998, *Carnivorous Plant Newsletter* 27(1): 16–17.

The formal epithet *luteola* is derived from the Latin *luteolus* (yellowish) and refers to the colouration of this form. Similar to *S. purpurea* ssp. *purpurea* fm. *heterophylla*, this form altogether lacks red colouration and consequently the leaves and flowers are pure yellowish green. Multiple strains of this form have reportedly been discovered in Florida and Alabama, although they are much more rare than the equivalent form of the northern subspecies. Other than colouration, this form is identical to the typical variety of *S. purpurea* ssp. *venosa* var. *burkii*. Note, however, that this is a red-pigment-free form of *S. purpurea* ssp. *venosa* var. *burkii*, not *S. purpurea* ssp. *venosa* as the colouration of the petals is white (reflecting *variety burkii*) rather than yellowish green which one would expect in the typical variety (as demonstrated in *S. purpurea* ssp. *purpurea* fm. *venosa*).

Sarracenia purpurea ssp. *venosa* var. *montana* Schnell and Determann

Original description: Schnell, D. E., and R. Determann, 1997, *Castanea* 62: 60.

The varietal name *montana* is derived from the Latin *montanus* (mountainous) and refers to the distribution of this variety in upland habitat in the southern Appalachian Mountains where the states of Georgia, North Carolina, South Carolina and Tennessee come together. The range of *S. purpurea* ssp. *venosa* var. *montana* is isolated from the rest of the territory of *S. purpurea* ssp. *venosa*. Much of the original habitat of this subspecies has been destroyed, and consequently it is highly endangered in the wild. *S. purpurea* ssp. *venosa* var. *montana* differs from the typical variety of this subspecies in that the lid is more conical in shape which results from the sides of the lid curling around the pitcher opening (Figure 213). From limited observations, it also appears that the tubular section of the leaf is generally more ventricose, and that the colouration of the foliage is more vibrant, than in the typical variety.

Figure 212 (facing page). The leaves of *Sarracenia purpurea* ssp. *venosa* var. *burkii* growing in northern Florida.

Figure 213. The leaves of *Sarracenia purpurea* ssp. *venosa* var. *montana*.

Sarracenia rubra Walter

Original description: Walter, T., 1788, *Flora Caroliniana, secundum*, 152.

The specific epithet *rubra* is derived from the Latin *rubra* (red) and refers to the colouration of the flowers and foliage.

The range of *S. rubra* consists of several disjunct populations distributed across the southeastern part of the United States, ranging from southern North Carolina through South Carolina, Georgia, Florida and Alabama to the southeastern edge of Mississippi (Figure 214). The fragmented range of this species has allowed the diversification and evolution of five distinct subspecies, but since each subspecies occurs only in its own, sometimes small, range, *S. rubra* is especially vulnerable to habitat loss and extinction. *S. rubra* ssp. *alabamensis* and *S. rubra* ssp. *jonesii* persist only at a handful of dwindling sites in central Alabama and on the border of North Carolina and South Carolina, respectively. Both subspecies face very uncertain futures and are listed as highly endangered in *CITES Appendix I* and on the *List of Endangered and Threatened Plants* of the United States.

S. rubra ssp. *gulfensis* occurs in the western Florida Panhandle and *S. rubra* ssp. *wherryi* is restricted to a small area in southwestern Alabama and southeastern Mississippi. These two subspecies are also increasingly rare and, in both cases, the number of known populations has halved within the last fifteen years. *S. rubra* ssp. *rubra* was originally distributed along the Atlantic Coastal Plain of North Carolina, South Carolina and eastern Georgia. While it is not critically endangered, the vast majority of its original habitat has already been destroyed and remaining populations are becoming increasingly scarce and isolated.

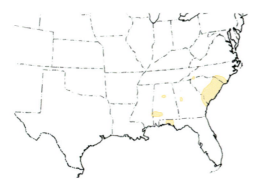

Figure 214. The range of *Sarracenia rubra*.

Figure 215. The flower of *Sarracenia rubra*.

A disjunct population of *S. rubra* spp. *rubra* occurs in western Georgia; these plants possess traits that are reminiscent of both *S. rubra* ssp. *gulfensis* and *S. rubra* ssp. *rubra* and are widely regarded as the ancestral midway form between those two subspecies.

The flower of *S. rubra* is typically 2–4 cm in diameter and borne on a 15–60-cm-tall flower scape. The rounded petals and slightly pointed sepals are crimson or maroon in colour (Figure 215). The umbrella-shaped pistil is reddish green. The morphology of the flower is roughly consistent throughout all subspecies of *S. rubra*.

The pitcher of *S. rubra* varies profoundly in terms of shape and size among the described subspecies, and indeed, with respect to pitcher morphology, *S. rubra* is the most variable species within the genus. Many infraspecific variants are discernable although none have yet been formally described or named.

Figure 216. The leaves of *Sarracenia rubra* ssp. *rubra.*

Sarracenia rubra ssp. *rubra*

The leaves of *S. rubra* ssp. *rubra* are erect, tubular, 10–35 cm tall and 1–2 cm wide. The lower and middle sections of the foliage are slender and tubular, but they broaden gently towards the pitcher opening (Figure 216). The ala is 2–10 mm wide, narrow below the pitcher opening and at the base of the leaf and broad at the midsection. The lid is elongated and terminates in a slight point. The foliage is predominantly yellowish green and lined with variable red veins. The upper surface of the lid is usually coppery yellow or, in some strains, dark bronze. *S. rubra* does not produce true ensiform noncarnivorous leaves, although in shaded habitat the ala is expanded and the tubular aspect of the foliage is reduced. The populations of *S. rubra* in western Georgia allegedly represent the form midway between *S. rubra* ssp. *rubra* and *S. rubra* ssp. *gulfensis* that was isolated as the *gulfensis* ancestor migrated southwards to the Gulf Coast. The lid is less elongated and the pitcher opening is more circular, similar to *S. rubra* ssp. *gulfensis*, but the size and shape of the leaf bears similarity to *S. rubra* ssp. *rubra*.

Sarracenia rubra ssp. *rubra* 'Green Variant' (Undescribed)

Various strains of *S. rubra* ssp. *rubra* have been discovered that are entirely free of red colouration and consequently the flowers and foliage appear pure yellowish green in colour. Other than the colouration, this variant is identical to the typical variety of this subspecies.

Sarracenia rubra ssp. *alabamensis* (Case and Case) Schnell

Original description: Schnell, D. E., 1978, *Castanea* 43: 260–261.

The subspecific epithet *alabamensis* refers to the state of Alabama where this subspecies is endemic. The pitchers of *S. rubra* ssp. *alabamensis* are erect, tubular and 15–45 cm tall and 2.5–4 cm wide (Figure 217). The lower parts of the leaf are slender, but they broaden gently towards the pitcher opening. The lid is broad and oval shaped although in some strains it is elongated and highly corrugated. The ala is 2–12 mm wide but less prominent than in the other *S. rubra* subspecies. In many strains, faint, white areoles are present below the pitcher opening and on the back of the lid. The leaves are typically greenish yellow and lined with variable red veining. Some strains produce vibrant golden-yellow coloured leaves, while others produce foliage that is flushed pastel pinkish red. It is difficult to document the full diversity of colour found in this subspecies since the vast majority of its wild populations have been destroyed. Many botanists treat *S. rubra* ssp. *alabamensis* as a separate species, *S. alamabensis* Case and Case.

Sarracenia rubra ssp. *gulfensis* Schnell

Original description: Schnell, D. E., 1979, *Castanea* 44: 217–223.

The subspecific epithet *gulfensis* refers to the Gulf Coast of Florida where this subspecies is endemic. The pitchers of *S. rubra* ssp. *gulfensis*

Figure 217 (facing page). The leaves of *Sarracenia rubra* ssp. *alabamensis*.

are 12–60 cm tall and proportionately stout (Figure 218). The leaf broadens close to the base so that the middle and upper sections are cylindrical. The pitcher opening is 2–4 cm wide. The lid is circular or oval shaped although greatly elongated in some strains. The ala is 2–10 mm wide, but narrow below the pitcher opening and broad at the midsection of the leaf. The pitchers are predominantly greenish yellow or orangey red in colour and lined with variable red veins, and often the underside of the lid suffuses pure red. In some strains, the leaves are uniform scarlet and lined with dark purple veins (Figure 218). The colouration of the foliage darkens over the course of the summer.

Sarracenia rubra ssp. *gulfensis* 'Green Variant' (Undescribed)

Various strains of *S. rubra* ssp. *gulfensis* have been discovered that are entirely free of red colouration and, consequently, the flowers and foliage appear pure yellowish green in colour. In some strains white areoles line the upper parts of the leaf, especially the lid and back of the pitcher opening. Other than the colouration, this variant is identical to the typical variety of this subspecies.

Figure 218. Leaves of *Sarracenia rubra* ssp. *gulfensis*, showing (left) typical colouration and (right) scarlet with dark purple veins.

Sarracenia rubra ssp. *jonesii* (Wherry) Wherry

Original description: Wherry, E., 1972, *Castanea 37*: 146.

The subspecific epithet *jonesii* honours the American botanist Dr. F. M. Jones who studied and examined many *Sarracenia* species during the early 20th Century. The pitchers of *S. rubra* ssp. *jonesii* are 15–50 cm tall and 1.5–3 cm wide at the pitcher opening (Figure 219). The lower and middle sections of the leaf are slender and narrow. The girth of the leaf broadens at the midsection but is constricted below the pitcher opening so that the upper section appears bulged. The ala is 2–10 mm wide.

Figure 219. The leaves of *Sarracenia rubra* ssp. *jonesii.*

The leaves are predominantly greenish yellow or orangey in colour and lined with variable red venation. Some strains exhibit exaggerated veins and appear predominantly red. It is difficult to evaluate what was the full natural diversity of this subspecies since the vast majority of wild populations have been destroyed.

Sarracenia rubra ssp. *jonesii* 'Green Variant' (Undescribed)

Multiple strains of *S. rubra* ssp. *jonesii* have been discovered that are entirely free of red colouration and, consequently, the flowers and foliage appear pure yellowish green in colour. Other than the colouration, this variant is identical to the typical variety of this subspecies.

Sarracenia rubra ssp. *wherryi* (Case and Case) Schnell

Original description: Schnell, D. E., 1978, *Castanea* 43: 260–261.

The subspecific epithet *wherryi* honours the American botanist Dr. E. T. Wherry who conducted various studies on *Sarracenia* species during the 1920s and 1930s. The pitchers of *S. rubra* ssp. *wherryi* are 12–30 cm tall and 1.5–3 cm wide at the pitcher opening (Figure 220). The exterior of the leaf is lined with 0.1–0.7-mm-long white hairs. The ala is 2–12 mm broad. Typically, the pitchers are coppery green or yellowish and lined with fine red veins. In most strains, the upper surface of the lid suffused coppery red over the course of summer. A population of giant *S. rubra* ssp. *wherryi* which produce 30–40-cm-tall leaves was discovered at Chatom, Alabama; these plants are known informally as the Chatom Giant Form.

Sarracenia rubra ssp. *wherryi* 'Green Variant' (Undescribed)

Various strains of *S. rubra* ssp. *ruba* have been discovered that are entirely free of red colouration and, consequently, the flowers and foliage appear pure yellowish green in colour. Other than the colouration, this variant is identical to the typical variety of this subspecies.

Sarracenia rubra ssp. *wherryi* 'Yellow Flowered Variant' (Undescribed)

In contrast to the usual red flowers, some strains of *S. rubra* have been discovered which produce pure yellowish-green or yellowish-orange

Figure 220. The leaves of *Sarracenia rubra* ssp. *wherryi.*

flowers but, in all other respects, are typical. The floral colouration is attributed to a faulty gene, but the gene is stable and will be inherited if two yellow-flowered strains are cross pollinated.

Sarracenia Hybrids

The distribution maps of *Sarracenia* species located throughout this chapter illustrate that the territories of several *Sarracenia* species overlap so that, in some areas, multiple species occur together. Similar to the situation with *Heliamphora*, where multiple species grow together, all of the flowers of all species present are pollinated by the same insects, mainly bees in the case of *Sarracenia*, and consequently hybrids are freely produced. Some *Sarracenia* species grow in isolation, so not all of the possible combinations of *Sarracenia* hybrids occur in the wild. The known natural *Sarracenia* hybrids are listed in Table 2.

All *Sarracenia* are inter-fertile and self-fertile, so an endless number of hybrids and complex hybrids, those involving two hybrids or one

Table 2. Names and Parentage of the Known Natural *Sarracenia* Hybrids.

BINOMIAL	PARENTAGE
S. x areolata	*S. alata x leucophylla*
S. x ahlesii	*S. alata x rubra*
S. x catesbaei	*S. flava x purpurea*
S. x chelsonii	*S. purpurea x rubra*
S. x courtii	*S. psittacena x purpurea*
S. x excellens	*S. leucophylla x minor*
S. x exornata	*S. alata x purpurea*
S. x formosa	*S. minor x psittacena*
S. x gilpini	*S. psittacina x rubra*
S. x harperi	*S. flava x minor*
S. x mitchelliana	*S. leucophylla x purpurea*
S. x moorei [a]	*S. flava x leucophylla*
S. x popei	*S. flava x rubra*
S. x readii	*S. leucophylla x rubra*
S. x rehderi	*S. minor x rubra*
S. x swaniana	*S. minor x purpurea*
S. x wrigleyana	*S. leucophylla x psittacina*
	S. flava x psittacina [b]

[a] This hybrid was previously named *S. x mooreana*.

[b] This hybrid is theoretically possible because the ranges of the species overlap, but it has not been formally reported.

species and one hybrid, can be produced. Complex hybrids do occur naturally, but they are relatively rare.

Although the differing characteristics of a hybrid's two parents are inherited and blended equally among the offspring, in *Sarracenia* red colouration of the leaves and flowers is always dominant over yellow or green colouration (Figure 221). For example, if a red-flowered species such as the typical variant of *S. leucophylla* hybridizes with a yellow-flowered species, such as *S. flava*, the offspring will have orange- or red-coloured flowers but not pure yellow ones. If the *S. flava x leucophylla* hybrid rebreeds with *S. flava*, the next generation of hybrids will still have orangey-red coloured flowers and, even after several additional generations of rebreeding with *S. flava*, the flowers will still be red tinted because of the dominance of the red colouration. However if *S. flava x leucophylla* hybridized with *S. leucophylla*, after just one or two generations the flower

Figure 221. A leaf of the hybrid *Sarracenia flava x leucophylla*, which is synonymous with the cultivar *Sarracenia x moorei*.

of the offspring will appear pure red and increasingly similar in colour to the flower of *S. leucophylla*.

The first hybrid produced in cultivation *(S. flava x leucophylla)* was bred and grown by Dr. David Moore and Wilhelm Keit at the Glasnevin Botanic Gardens of Dublin, Ireland, in 1870. During the late 19th and early 20th centuries, various horticultural hybrids of *Sarracenia* were created by different European nurseries and institutions. Many hybrids were given binomial names which could be used as an alterative to listing the parentage of the hybrid, for example, *S. flava x leucophylla* was named *S. x moorei* after Dr. Moore. Unfortunately, the parentage of the binominal names was not always recorded and the same binomial names were given, sometimes accidentally, to the multiple hybrids. Eventually, the system became complicated and largely abandoned. To avoid these problems, hybrids now are identified by their parentage (Table 3).

Table 3. Binomial Names and Parentage of Some Common Horticultural *Sarracenia* Hybrids.

BINOMIAL[a]	PARENTAGE
S. x caroli-schmidtii	*S. [purpurea x rubra] x purpurea*
S. x cookiana	*S. [flava x minor] x purpurea*
S. x diesneriana	*S. [psittacina x purpurea] x flava*
S. x exornata	*S. purpurea x [flava x minor]*
S. x exculta	*S. purpurea x [flava x minor]*
S. x exsculpta	*S. purpurea x [flava x minor]*
S. x georgiana	*S. [purpurea x [purpurea x rubra]] x [minor x purpurea]*
S. x illustrata	*S. alata x [flava x purpurea]*
S. x illustrata	*S. [flava x purpurea] x flava*
S. x kaufmanniana	*S. [purpurea x rubra] x purpurea*
S. x laschkei	*S. [purpurea x psittacina] x [flava x leucophylla]*
S. x melanorhoda	*S. [flava x purpurea] x purpurea*
S. x milata	*S. minor x alata*
S. x miniata	*S. minor x alata*
S. x mixta	*S. [alata x oreophila] x leucophylla*
S. x moorei [b]	*S. flava x leucophylla*
S. x sanderae	*S. leucophylla x [[flava x minor] x purpurea]*
S. x sanderiana	*S. leucophylla x [leucophylla x rubra]*
S. x schoenbrunnensis	*S. [purpurea x psittacina] x [[flava x minor] x purpurea]*
S. x superba	*S. [leucophylla x minor] x leucophylla*
S. x umlauftiana	*S. [purpurea x psittacina] x [leucophylla x psittacina]*
S. x vetteriana	*S. [flava x [flava x purpurea]] x [flava x purpurea]*
S. x vittata	*S. purpurea x [purpurea x rubra]*
S. x vittata-maculata	*S. purpurea x [purpurea x rubra]*
S. x vogeliana	*S. [purpurea x psittacina] x [flava x purpurea]*
S. x westphalii	*S. [[leucophylla x purpurea] x [flava x leucophylla]] x leucophylla*
S. x willisii	*S. [purpurea x psittacina] x [[purpurea x flava] x purpurea]*
S. x willmottiae	*S. [flava x purpurea] x purpurea*

[a] This list is provided for illustrative and reference purposes only; these names are no longer used.

[b] This hybrid was previously named *S. x mooreana*.

Selected *Sarracenia* Cultivars

Cultivars are not "varieties" of plants in the taxonomic sense, but rather are specific strains that exhibit worthy characteristics that are maintained and thereby preserved through the process of cultivation. The majority of cultivars are bred and enhanced horticulturally through selective breeding techniques which allow characteristics such as vigor, colour and size to be enhanced to create plants with unique traits that would not arise under normal breeding conditions. Most *Sarracenia* cultivars are produced by amateur enthusiasts and commercial nurseries in attempts to create ever more unusual and interesting stock. Cultivars can be hybrids as well as species and many of the most striking and unusual *Sarracenia* cultivars result from complex crossbreeding programmes involving combinations of species that could not occur naturally.

Once bred and selected, each *Sarracenia* cultivar is submitted for registration through the International Carnivorous Plant Society (see *www.carnivorousplants.org*). The details of the cultivar's parentage are recorded and the cultivar is given a unique name that enables it to be identified with ease and confidence. Since cultivars are particular horticultural strains rather than taxonomic varieties, all cultivars can only be reproduced asexually through cuttings or division since a fertile seed of a cultivar would never yield offspring identical to either of its parents. Below are described eleven long established cultivars that have been widely distributed across most of the world and represent some of the most beautiful *Sarracenia* under cultivation. Many of the following cultivars were developed before the modern system of registration was instituted and therefore not all have been formally described.

As with all *Sarracenia*, the following cultivars require exposure to several hours of direct sunlight each day during the growing season in order to develop their spectacular and unique colouration. Insufficient light levels will cause the foliage to appear green and etiolated and the unique characteristics will be suppressed.

Sarracenia 'Brooks Hybrid'

S. 'Brooks Hybrid' is an exceptionally large and vigorous strain of *S. flava x leucophylla*. The pitchers frequently grow taller than 100 cm in height and are particularly robust and vigorous. The leaves are predominantly

Figure 222. A leaf of *Sarracenia* 'Brooks Hybrid.'

yellowish green, the lid is lightly dappled with white areoles and the back of the pitcher opening develops delicate purple colouration and veins (Figure 222). This excellent cultivar was bred by Mr. M. Brooks during the 1980s.

Sarracenia 'Burgundy'

The beautiful *S.* 'Burgundy' strain of *S. flava* var. *rubricorpora* was selected and named by Adrian Slack at his renowned carnivorous plant nursery, Marston Exotics, in Hereford, United Kingdom. The 50–70-cm-tall leaves develop exceptionally vibrant scarlet colouration which contrasts with an unusual purplish-blue sheen that is discernable mainly in the lower parts of the leaves (Figure 223). The pitcher opening and lid are vivid yellow and lined with prominent dark red veins and purple

Figure 223. The leaves of *Sarracenia* 'Burgandy.'

rugelii-style colouration. *S.* 'Burgundy' is truly among the most magnificent of all cultivated strains of *Sarracenia*.

Sarracenia 'Diane Whittaker'

S. 'Diane Whittaker' consists of the parentage *S. (leucophylla x minor) x leucophylla* and was created by Adrian Slack during the 1980s at Marston Exotics and later selected and named by Mike King of Shropshire Sarracenias nursery. The 30–45-cm-tall pitchers are predominantly yellowish green in colour. The upper section of the leaf is mottled with white areoles and delicate red veins (Figure 224). Over the course of autumn and winter, the leaves flush a beautiful shade fuchsia pink so that the foliage looks most magnificent at Christmas time.

Figure 224. A leaf of *Sarracenia* 'Diane Whittaker.'

Sarracenia 'God's Gift'

S. 'God's Gift,' similar to S. 'Diane Whittaker,' originated from Adrian Slack's breeding programme but was later selected and named by Mike King and distributed primarily through Shropshire Sarracenias. The foliage of S. 'God's Gift' is predominantly yellowish green and dappled with white areolation and delicate red-and-purple veins (Figure 225). This cultivar is noted for its vigour and rapid growth.

Sarracenia 'Judith Hindle'

S. 'Judith Hindle' was created and named by British *Sarracenia* breeder Alan Hindle during the 1990s. Its parentage is S. *leucophylla x (flava x purpurea)*. S. 'Judith Hindle' produces 20–30-cm-tall leaves that are predominantly red and dappled with variable white areolation that is netted with intricately bright red veins. Over the course of summer, the

Figure 225. The leaves of *Sarracenia* 'God's Gift.'

areolation tints attractive shades of purple and pink and, usually, the leaf suffuses pure crimson by autumn. In recent years, *S.* 'Judith Hindle' has been tissue cultured and distributed worldwide.

Sarracenia 'Juthatip Soper'

S. 'Juthatip Soper' was produced by Mathew Soper of Hampshire Carnivorous Plants and has the parentage *S. x (leucophylla x purpurea) x leucophylla*. This beautiful cross produces 30–40-cm-tall pitchers that are heavily dappled with pink-tinted areolation and delicate red veins that suffuse a more intense shade of red with age (Figure 226). The flowers of this handsome cultivar are pure red. *S.* 'Juthatip Soper' has won several prominent awards in the United Kingdom, including the prestigious Royal Horticultural Society's Award of Merit.

Figure 226. A leaf of *Sarracenia* 'Juthatip Soper.'

Figure 227. The leaves of *Sarracenia* 'Marston Clone.'

Sarracenia 'Marston Clone'

S. 'Marston Clone,' a truly outstanding strain of *S. flava x leucophylla*, was created and selected by Adrian Slack in the 1980s and retailed for many years as his choice *Sarracenia* hybrid through his nursery Marston Exotics. It produces elegant 60–70-cm-tall pitchers which are predominantly yellow, light green and white in colour and ornately lined with a small amount of red and pink veining (Figure 227). Its flowers are noted as being an attractive shade of orangey red. It is particularly vigorous and relatively fast growing.

Sarracenia 'Maxima'

S. 'Maxima,' also developed and selected by Adrian Slack at Marston Exotics, is a particularly vigorous and tall growing strain of *S. flava* var. *flava* (Figure 174) that produces robust leaves in excess of 100 cm tall. The noncarnivorous leaves are noted for being slightly blue tinted. *S.* 'Maxima' is extremely vigorous and widely acclaimed as the most beautiful and resilient strain of *S. flava* var. *flava* in cultivation

Sarracenia 'Sarracenia Nurseries Select'

S. 'Sarracenia Nurseries Select' is a spectacular giant strain of *S. flava x leucophylla* that was produced and selected by breeders at Sarracenia Nurseries in England during the 1980s. The upper parts of the leaf develop intense red and crimson colouration and are dominated by intensely dark red veins which contrast with subtle white and pink areolation that lightly dapples parts of the lid and pitcher opening (Figure 228). *S.* 'Sarracenia Nurseries Select' is extremely vigorous and readily produces leaves in excess of 80 cm in height.

Sarracenia 'Schnell's Ghost'

S. 'Schnell's Ghost' is a particularly beautiful, pale strain of *S. leucophylla* that lacks obvious red venation (Figure 229). Red pigment is not altogether absent and is occasionally apparent on aging or dying leaves, but for the most part, the foliage of *S.* 'Schnell's Ghost' appears white and green. This cultivar has been grown for many decades and apparently was collected from the wild by Dr. Donald Schnell in 1972. Wilson (2001) records in detail the complicated history that surrounds this plant prior to its being named, and notes that Schnell originally described the plant as very pale and "ghost-like," and so it came to be

Figure 228. A leaf of *Sarracenia* 'Sarracenia Nurseries Select.'

Figure 229. A leaf of *Sarracenia* 'Schnell's Ghost.'

Figure 230. The unique flower of *Sarracenia* 'Tarnok,' with typical leaves in the background.

named, affectionately, in his honour. The most unusual characteristic of *S.* 'Schnell's Ghost' is that it produces a pure yellow flower rather than the crimson one that is typical of *S. leucophylla*. *S.* 'Schnell's Ghost' is relatively vigorous and grows quickly.

Sarracenia 'Tarnok'

S. 'Tarnok' is an unusual strain of *S. leucophylla* which produces flowers that are uniquely mutated and formed of multiple concentric whorls of tepals (Figure 230). As a result, this unusual structure causes the flower to display a conspicuous "pompom" shape rather than the regular *Sarracenia* structure. Determann and Groves (1993) record that this unusual floral characteristic was originally discovered during the 1970s by Mr. Coleman Tarnok in a population of *S. leucophylla* growing in the Perdido region of Baldwin County, southern Alabama, and that a strain of these unusual plants entered cultivation and were later named in Mr. Tarnok's honour. According to Determann and Groves (1993), the tepals are predominantly green in colour when new, but gradually change to a deep maroon which is retained throughout the growing season and often as late as midwinter. In all respects other than the flower, *S.* 'Tarnok' is typical of *S. leucophylla*. Due to the floral mutation, this cultivar is believed to be sterile as all attempts to pollinate individuals have failed to yield seed.

Habitat Loss and the Threat of Extinction

Habitat change is proceeding at an unprecedented rate and scale throughout the world as the human population increases and the power of technology magnifies the human potential for altering and shaping the global environment. As a result, natural ecosystems of all sizes and complexities are undergoing change, often disruptive or destructive change, and one profound consequence is that many biological organisms are becoming endangered or, in all too many cases, rendered extinct. The American pitcher plants are among those organisms that are affected by the process of environmental change and several species are threatened with the imminent risk of extinction. Unsurprisingly, the five genera of American pitcher plants are threatened very differently owing to their dissimilar distribution and ecology.

The South American genera of pitcher plants — *Brocchinia*, *Catopsis* and *Heliamphora* — mostly occur in remote, inaccessible, sparsely populated and lightly industrialized regions of the continent and have been little affected by the activities of humans. In contrast, the North American genera *Darlingtonia* and *Sarracenia* have become increasingly threatened (Figure 231), and in some cases endangered, with the growth of the North American population and economy. *Catopsis* in Middle America faces growing pressures from increasing human populations and the changes in land cover and use that accompany the presence of more people.

Figure 231 (facing page). *Sarracenia rubra* ssp. *rubra* — currently listed in *CITES Appendix II* — is among the *Sarracenia* taxa that are increasingly threatened with extinction.

Remoteness and inaccessibility are the main factors that have protected the highland species of South American pitcher plants from human disturbance. Populations of *Brocchinia* and *Heliamphora* that occur on the lofty summits of the tepuis have been known to the outside world for only a short time, and they continue to be so far detached from human societies that, for the most part, they remain in what is virtually pristine, natural habitat. Although the presence and impact of tourists in the Guiana Highlands are certainly increasing, the overwhelming majority of the tepuis are nearly inaccessible and seldom visited. Venezuelan authorities are playing an increasingly responsible role in regulating tourism to the tepuis in general and minimizing the potentially negative impact of this economy on the region. Even on Mount Roraima where tepui-tourism is concentrated, the effect of visitors is relatively minimal and restricted to a very small proportion of the total surface area of the plateau summit. The long-term outlook for highland *Brocchinia* and *Heliamphora* populations is optimistic, especially since the majority of the tepuis now lie within vast, newly established national parks of southern Venezuela, northern Brazil and Guyana where mining and construction activities are strictly controlled and all dangers to wildlife are stringently managed. The only significant foreseeable threats to highland pitcher plant populations in South America are the accidental introductions of non-endemic biota, such as seeds of exotic plants inadvertently transported to the region on the clothing of visitors, or long-term, wide-scale climate change which may alter rainfall patterns and microclimatic zonation of the tepui habitats. Both threats are increasingly being monitored by Venezuelan ecologists and do not appear to represent immediate dangers.

The near-future outlook for lowland populations of pitcher plants in the Guiana Highlands is similarly optimistic. Lowland *Brocchinia*, *Catopsis* and *Heliamphora* populations mostly occur on land with marginal economic value that is located far from large human settlements, and consequently they remain essentially undisturbed. Over the next century, rising South American population densities may increase land-use pressure and the demand for water resources could potentially threaten lowland pitcher plant habitats, but since the majority of the pitcher plant populations and habitats occur within Venezuela's national parks, most notably Canaima National Park, the longer term outlook for these plants appears relatively secure.

Although relatively secure within the Guiana Highlands, lowland populations of *C. berteroniana* are arguably more vulnerable to disruption because of habitat change that is taking place in other parts of their range. *C. berteroniana* depends directly upon forested habit for its survival, and that habitat is in swift decline across many parts of the species' range. The loss of forested areas is especially acute in Central America and the Caribbean where urban expansion, the clearance of land for agriculture and the growth of the fuel-wood industry continue to take place on an ever-increasing scale. Unfortunately, secondary forests require several decades of recovery before they provide suitable habitat for *C. berteroniana* and, as a result, widespread habitat disruption usually causes a long-lasting crash in the numbers of *C. berteroniana* and other bromeliad species. In southern Florida, much of the original habitat of *C. berteroniana* was lost during the second half of the 20th Century and most of the remaining populations of *C. berteroniana* persist only in nature reserves, most notably the Everglades. The introduction of insect pests also represents a very real threat to *C. berteroniana* populations; the presence of the Mexican bromeliad weevil (*Metamasius callizona*) in Florida, for example, has devastated local populations of *Catopsis* and other bromeliads. The overall status of *C. berteroniana* is unclear in many parts of its range, most notably Colombia and the south of Venezula, but populations are known to occur in nature reserves in most countries throughout its vast geographic range and, therefore, the long-term outlook for survival is relatively positive, at least in protected areas. A more robust inventory and extensive monitoring of populations of this species would be helpful in developing a more realistic assessment of its current, and probable near-future, status.

The conservational status of the North American pitcher plants is unfortunately much more critical, especially that of several *Sarracenia* species. Groves (1993) estimates that only 2.5% of the *Sarracenia* habitat that was available at the onset of European colonization currently remains in the southeastern United States (Figure 232), and indeed this figure seems realistic considering that colonial-era explorers such as Mark Catesby recorded vast wetland habitats that stretched for hundreds of

Figure 232 (following pages). A small population of *Sarracenia alata* glowing in the early morning sunlight in eastern Louisiana.

kilometers across the landscapes of the region. Today, only a few scattered and isolated remnant populations of *Sarracenia* occur throughout this region, and the majority of those survive only in protected nature reserves. This situation is especially dire when one considers that the entire range of seven of the eight species of *Sarracenia* fall entirely within this region (Figure 233). Appreciably less of the habitat of the eighth species, *S. purpurea*, has been destroyed. Increasingly, however, the habitat of the southern subspecies *S. purpurea* ssp. *venosa* is being lost and in many counties within its historic range along the Gulf Coast, it is now locally extinct.

Three main factors have contributed to the wide-scale degradation and destruction of *Sarracenia* habitat in Canada and the United States. These are:

1. **Artificial drainage of wetlands.** The growth of the North American agricultural industry over the past three hundred years often has depended on the conversion of wetlands to fertile rural cropland or pastures through the artificial modification of natural drainage systems, including surface and sub-surface flow as well as the flow of streams. Artificial drainage channels in particular (Figure 234) have effectively reduced the local water table and displaced wetland habitat and the species that they supported, including *Sarracenia*.

2. **Fire suppression.** During the past century the prevalence of wildfires has been artificially suppressed across rural areas of Canada and the United States in order to protect property and agricultural livelihoods. For the most part, wetland habitats naturally depend upon wildfires to periodically reduce undergrowth and forest cover and maintain an equilibrium between herbaceous, low growing plant species and dominant forest cover. Long-term wild-fire suppression has directly affected ecological succession and enabled longer-lived, closed-canopy vegetation to become dominant which has directly caused the displacement of many wetland species.

3. **Commercial tree farming.** The advent of wide-scale commercial forestry plantations and subtropical tree-crop groves

Figure 233. A meadow in southern Alabama with a mixed population of *Sarracenia flava* and *Sarracenia leucophylla*.

during the past century has destroyed or degraded large areas of North American wetlands and reduced or displaced the plants and animals that inhabited these environments.

Urban expansion, highway and other transportation-related construction and the use of fertilizer, pesticides and herbicides have also contributed to the dramatic reduction of the *Sarracenia* range. In fact, habitat concerns and population viability for all species in the genus are so acute that all are listed in *CITES*. *S. oreophila*, *S. rubra* ssp. *alabamensis* and *S. rubra* ssp. *jonesii* are considered imminently threatened with extinction and are therefore included in *CITES Appendix I*, and the other four species are listed as potentially threatened and included in *CITES Appendix II*.

The demise of *Sarracenia* habitat (Figure 235) has occurred with particular speed during the past few decades. Mr. Jim Miller of Tallahassee, Florida, has recorded in detail the loss of *Sarracenia* populations in the

Figure 234. The artificial draining of freshwater wetlands began almost as soon as as the European settlement of North America began, and it has continued vigorously to thepresent.

southeastern United States during this time and has kindly provided the following account of his personal observations of this phenomenon.

> *As late as 1980, you could stand on the Alabama side of several major highways that connect that state with Florida and see hundreds of thousands of* S. leucophylla *in fields that stretched as far as the eye could see. At that time I regularly travelled across the Gulf Coast in search of carnivorous plant populations to study and observe. One of the most dramatic experiences in the field took place just after I moved to Tallahassee from Miami in June of 1977. During the 4th of July holiday weekend, I left my apartment in the early hours of the morning and arrived right at dawn on the Alabama side of Highway 90. The sight of so many pitcher plants illuminated by the golden light of the rising sun was quite literally breathtaking. In fact, it was so overwhelming that I failed to grab my Nikon camera. I just stood and stared and smiled.*
>
> *At that time, I never considered just how fleeting such scenes would be. In April of 2005, Stewart McPherson, Brooks Garcia and I drove that route. At what I judged to be the exact spot where once so many pitcher plants had so grandly stood, I looked out of the window at the*

surrounding fields and saw not a single plant anywhere. I cannot possibly relate to you how my heart ached.

In 1971, I made my very first field trip into the heart of the Green Swamp in the southeast corner of North Carolina. The Green Swamp came into being many tens of thousands of years ago when a meteor shower struck that part of the North American continent and indented the topography and allowed the formation of extensive wetlands. Satellite imagery in the late 1970s identified the jagged outline of that ancient impact. It was in the Green Swamp where I saw my very first Venus Flytrap in the wild; not just a single plant, but hundreds of thousands of them so numerous that they formed a dense mat on the ground that even hearty weeds could not compete with.

In 1984, I returned to the same location and gazed at hundreds upon hundreds of acres of dead and dying carnivorous plants. Deep drainage ditches had been cut, crisscrossing this once pristine habitat, to lower the local water table. The ground had been disked and scraped so that pines and sweet gums could be grown and eventually commercially

Figure 235. Populations of many species of pitcher plants, such as this field of *Sarracenia leucophylla* in southern Alabama, are disappearing as a direct result of habitat destruction and commercial collecting.

harvested. We searched for any Sarracenia *that might have survived. But after two hours in the hot summer sun, we gave up. The decimation was on a scale so great that it defied comprehension. All we found were thousands of dead pitcher plants, some scattered about and others pushed into piles with other foliage for eventual burning. If you drive that road today, you might even miss the small sign that announces that you have arrived at what's left of the great Green Swamp. Only a tiny percentage of the original habitat has been placed under the protection of the US Nature Conservancy. But gone are the field of Flytraps. Gone is the slow moving stream where I first discovered a huge, aquatic specimen of* Drosera intermedia. *Gone is the immense, spring-fed bog where* S. flava *once grew so tall and so thick that it was virtually impenetrable.*

Shall I go on? Maybe I would tell you about a beautiful location not far off Interstate 10 in the Florida Panhandle. There, near the town of DeFuniak Springs, was the most fantastic stand of S. flava *and many hybrids and backcrosses with* S. purpurea. *Indeed there were so many variations in form and color that it was clear that, despite its proximity to a main north-south highway, that population had been left untouched for many decades. The site is now gone. In its place is a Kentucky Fried Chicken restaurant. One week the plants were there and thriving. Three weeks later there was a parking lot.*

One final example to illustrate my point; just south of Interstate 10 in the Florida Panhandle, near Crestview, in 1979 there used to be a superb habitat for several species of carnivorous plants. A radio tower marked the location. If you walked just a short distance beyond the tower, you found yourself at a spring-fed pond. Along its banks and well into the deeper woods along the seeps were thick mounds of sphagnum moss. Even in the middle of summer, the sphagnum was always cool to the touch, perhaps as a result of the chilly spring waters in which it grew. In many places, growing on the tops of the moss were huge mats of Pinguicula primuliflora *(a carnivorous butterwort) and in the more open areas there were uniform stands of* S. rubra *ssp.* gulfensis. *On the south edge of the pond, there was a small field where* S. rubra *ssp.* gulfensis *grew along with* S. leucophylla.

Over the last two decades, the once sleepy town of Crestview has grown into a sprawling city. What was once a magical glade is now just a few hundred feet from almost continuous traffic on a four-lane highway. The site has a new owner who has dug extensive pits and trenches to drain the land and to change the vegetation. On the northeast corner of the property yet another strip mall now stands. In an effort to preserve and retain the diversity of the plants of that location, myself and other Sarracenia *enthusiasts collected some of the remnant plants with the owner's permission. His intention was to develop that entire piece of property and he expressed no remorse for the loss of this habitat. "Tell your friends to bring their trucks and take whatever you want," he said, waving a hand in casual dismissal of the site. "Load up everything you can find because in a week there won't be anything here."*

Each time I now travel to these areas, my heart is broken. I would estimate that less than five percent of the habitats I explored in the 1970s through the late 1990s still exist and support populations of carnivorous plants today. That number shrinks every day and there is no end in sight. We have lost the fight to protect these fragile ecosystems well before it ever became evident that a fight was needed to preserve them.

Some of the last Sarracenia *populations are preserved in wildlife reserves such the Apalachicola National Forest in Florida and the DeSoto National Forest in Mississippi. These populations are federally protected and the wetland habitat is well maintained and regularly burned. While these populations will likely always be maintained, they represent only tiny fragmented specks of the original* Sarracenia *populations, now only a memory captured on pieces of Kodachrome film.*

Perhaps the greatest tragedy of this situation lies in that the destruction of these wetland habitats is (at least to some degree) avoidable. Since 1975, I have visited once vast carnivorous plant populations distributed along the famous Yellow River in Florida. On the southern edge nearest the river several Sarracenia *spp. once grew, including* S. rubra *ssp.* gulfensis, S. leucophylla *and* S. psittacina *as well as an incredible diversity of hybrids. On the north side of the site, on slightly higher ground, there grew* S. flava. *Oddly, at this population there was a clear line of demarcation between these two groups of species. There were many other rare plants here; large specimens of* Pinguicula planifolia *grew in the wettest, muddiest areas along with huge colonies of* Drosera intermedia *and several species of* Utricularia. *A series of gigantic high-voltage power lines crossed this location. During the 1970s the vegetation beneath the lines and pylons was manually mowed to allow the power company access and to reduce the risk of wild fires. This practice was actually beneficial to the wetland ecosystem and, in particular, the carnivorous plants since it effectively controlled the dominant vegetation, namely woody shrubs, and allowed a diverse array of smaller plants to flourish. Today, mowing has been replaced by sprayed applications of herbicide, which from my experience kill carnivorous plants and possibly some species of terrestrial orchids. The broken ecosystem then becomes dominated by the thick re-growth of weeds and shrubs which dominate until the next herbicide spraying.*

In an effort to record the historic range and last remaining pristine populations of Sarracenia *spp., I began to film wild* Sarracenia *spp. populations in 1984 when the very first big and clunky video cameras became available. Today, with professional equipment, at every opportunity I film and document footage of pitcher plants, sundews and other carnivorous species as well as orchids and other rare plants in an effort to create and preserve digital records of the plants which I so love. With material dating back as far as 1971, I am creating and distributing records to serve as a resource in the future conservation and understanding of these increasingly rare plants. I regret that it was not possible to document the truly massive, original* Sarracenia *populations such as those of Yellow River in 1975 or the great Green Swamp in 1977. In a perfect world, I would not have to shoot video to protect the beauty of these magnificent habitats. I wish they would always remain so that I — or anyone else who might wish to do so — could return in the future and enjoy their natural beauty and splendour. It has taken so many millions of years for these improbable plants to evolve, yet their numbers have been decimated in just a few decades.*

If we can save just a tiny piece of what is left, we will be giving to our children an invaluable gift of natural beauty. History is being written every minute and every hour, the future of these plants and these beautiful natural habitats is not set. Let us all hope that this story will have a happy ending.

Jim Miller
October 28, 2005

The personal dedication and commitment of individuals such as Jim Miller certainly brings hope to the cause of conservation. For years, Jim has dedicated himself to the concept of producing a series of video records that will eventually serve as a "time capsule" that preserves rare and unique footage of natural *Sarracenia* populations and their extraordinary diversity. Through visually documenting these plants in their natural setting, it is Jim's hope that his efforts will be useful in future conservation initiatives, especially in efforts to re-establish populations of *Sarracenia* species. To further advance his objectives of preserving and sharing knowledge about carnivorous plants, Jim has established and will maintain an online resource site that will provide information about his projects and initiatives as they evolve. This site presently is located at *www.redfernnaturalhistory.com/sarraceniaarchives.htm.*

The disastrous loss of *Sarracenia* habitat has been compounded during recent decades by the devastating effect of illegal commercial poaching of *Sarracenia* plants used to supply the horticultural industry's demand for carnivorous plants. Although Robbins (1998) notes that the emergence of artificial horticultural techniques have reduced commercial harvesting of naturally occurring *Sarracenia* plants for the horticultural industry, illegal poaching by amateaur enthusiasts remains high and particularly significant with regards to the imperilled *Sarracenia* species listed in *CITES Appendix I*.

The impact of commercial collecting of the leaves of *Sarracenia*, in particular those of *S. leucophylla* for the floral industry, also remains a significant threat to wild populations (Figure 236). The collection of *Sarracenia* leaves is potentially a sustainable activity but a massive overexploitation of wild *Sarracenia* populations is occurring in many southern states and that is jeopardizing some populations of the *CITES Appendix II Sarracenia* species. In the spring of 2005, I revisited a number of large, well documented *S. leucophylla* populations in Alabama. The previous year, I had observed populations consisting of several thousand plants at both locations, but when I returned in 2005, both *Sarracenia* populations had been entirely stripped of leaves and only damaged and small *S. leucophylla* plants remained intact. The leaves of the mature plants had clearly been harvested as the stems were "clean cut" and human footprints across both populations were evident. Robbins (1998) notes that the damage caused to the soil and hydrology of boglands by the commercial collection of *Sarracenia* leaves is potentially more damaging to *Sarracenia* populations than is the direct collection of the leaves. The sensitivity of wetland substrate renders it highly vulnerable to erosion which can directly lead to large scale and permanent hydrological and ecological change. The continuing large-scale collection of *Sarracenia* leaves therefore affects *Sarracenia* populations in at least three direct ways:

1. It directly places stress on individual plants and reduces their productivity — which in the context of an entire population will profoundly influence the rate of flowering and population sustainability.

2. It directly influences substrate activity, including hydrological and erosive processes which, in turn, could have potentially significant local consequences on the physical and ecological stability of the site.

3. It directly reduces the long-term productivity of the population by removing organic matter from the site, which in turn depletes the nutrient pool of the ecosystem and thereby stresses the viability of the *Sarracenia* populations.

In the context of global climate change and a potentially warmer and drier environment in some parts of the world, the importance of conservation programmes aimed specifically at the preservation of wetland habitats and their constituent species are essential for the long-term survival of all *Sarracenia* species. The combination of natural and cultural processes — such as extended drought and the systematic suppression of wildfire or the increased harvesting of pitcher leaves — has exacerbated the rarity of many *Sarracenia* species. Several of the *CITES Appendix II* species, especially *S. rubra* ssp. *gulfensis, S. rubra* ssp. *wherryi, S. purpurea* ssp. *purpurea* fm. *heterophylla* and *S. purpurea* ssp. *venosa* var. *montana,* are becoming increasingly rare and their protective status is increasingly in need of re-evaluation.

S. purpurea is the only species of *Sarracenia* which remains widely distributed and relatively populous across much of its original range. There does, however, exist a distinct difference between the status of the two subspecies. The lack of disturbance across much of its vast territory, especially in remote areas of Canada, has enabled *S. purpurea* ssp. *purpurea* populations to persist in relative abundance throughout much of the species' former range. *S. purpurea* ssp. *venosa* and *S. purpurea* ssp. *venosa* var. *burkii,*

Figure 236 (facing page). The elegant leaves of *Sarracenia leucophylla* are often collected for use in the floral industry.

however, are considerably less abundant in the wild today than formerly, and *S. purpurea* ssp. *venosa* var. *montana* is considered to be severely threatened.

The responsible management of the remaining North American wetland habitats is of critical importance for the survival of *Sarracenia* populations and those of other highly threatened wetland plant and animal species. In recent decades, large and well maintained nature reserves have been established throughout Canada and the United States. The Okefenokee Wildlife Refuge in Georgia, the Green Swamp Preserve in North Carolina, The Blackwater River State Forest and the Apalachicola National Forest in Florida represent several important remaining wetlands where *Sarracenia* populations persist. All of these reserves are regularly and intentionally burned under fire-management programmes that approximate natural wildfires and the consequential ecological processes of fuel reduction, accelerated nutrient cycling, vegetation clearing and biotic succession to occur. In the case of these representative parks, the use of fire is an important part of maintaining healthy wetland ecosystems. The long-term sustainability of *Sarracenia* populations within these reserves is likely and, indeed, the survival of *CITES Appendix II Sarracenia* species is secure at least in managed nature reserves.

The long-term viability of the *CITES Appendix I Sarracenia* species is, however, much more questionable. All three endangered taxa currently occur at a very small number of sites and the total wild population of each is extremely small. Although the remaining habitat where these plants grow is secure, the small number of individual plants at each site and the consequent lack of significant genetic diversity represent major challenges to their survival. The Atlanta Botanical Gardens in Atlanta, Georgia, has engaged in a very successful conservation programme that aims to manage and restore habitat of the *CITES Appendix I Sarracenia* species and regularly manages the artificial burning of several habitats containing endangered *Sarracenia* taxa. This initiative represents an extremely important undertaking that will assist greatly with the survival of endangered *Sarracenia*. This programme is more fully explained at *www.atlantabotanicalgarden.org.*

An equally important project has been undertaken by the Meadowview Biological Research Station, Woodford, Virginia, primarily to preserve and restore habitat and populations of *CITES Appendix II Sarracenia* species with the objective of re-establishing the natural range of *Sarracenia* species that have become locally extinct in Virginia and Maryland. This conservation programme is directed towards the long-term sustainable management of *Sarracenia* habitat and is directly responsible for the ongoing preservation of multiple wetland areas. See *www.pitcherplant.org* for an overview of the Meadowview conservation programme.

A number of extremely worthwhile nonprofit organizations dedicated to the preservation of the wetland homes of *Sarracenia* and other carnivorous plants in the United States are identified at *www.redfernnaturalhistory.com/sarraceniaconservation.htm*. Conservation initiatives such as those identified at this site pose viable means of insuring that these beautiful plants are saved and protected for future generations. Proceeds from the sale of this book will be donated to help the conservation cause (Figure 237).

The status of *D. californica* is much more secure, and the outlook for future viability is considerably more optimistic. Because the range of *Darlingtonia* is primarily in mountains and along a rugged coast, *D. californica* has not been subjected to the same level of urban, agricultural, and transportation development pressures as has *Sarracenia*. In addition, a significant proportion of the original range and habitat of this species has been protected through well-managed conservation initiatives, and much of the beautiful coastline of northern California and Oregon has been set aside and protected as permanent nature reserves. Indeed, immediately north of Florence, Oregon, there is a wayside preserve that has been set up explicitly to protect *D. californica* — Oregon's only state park dedicated to the preservation of one particular plant species. While it is easy to see *D. californica* in the wild and the state park officers of California and Oregon generally help naturalists find and enjoy wildlife, it is important to remember that the collection of these plants from wild populations is strictly prohibited by law and must be respected by all visitors.

The threat of extinction facing the pitcher plants of the Americas is indeed very unequal. In 2006, none of the South American pitcher plants are listed in *CITES* whereas all of the North American species, with the exception of *D. californica*, are regarded as potentially or imminently endangered. This difference between the two groups is in many ways surprising since it is the South American genera, in particular the *Heliamphora*, that mostly occur across small, geographically restricted ranges often consisting of a single mountain or valley. In contrast, most species of *Sarracenia* were originally distributed across vast

Figure 237. The leaves of *Sarracenia alata* 'Red Lid Variant' represent part of the wide spectrum of variation in this species that has not yet been defined by taxonomists. It is crucial that the diversity within *Sarracenia* be appreciated before the few remaining large populations of species within the genus are destroyed.

areas of wetlands that collectively encompassed hundreds of thousands of square kilometers. This unequal threat emphasises the importance of the interplay of socioeconomic factors in the survival of species and habitats, and the crucial importance of awareness and direct action for the future conservation of the biodiversity of our world.

Several international organizations are working towards insuring a more secure future for the American pitcher plants. The Royal Botanic Gardens of Kew, through the launch of its Millennium Seed Bank Project, has included in its vast seed reserves seed stock of many carnivorous plant species, including American pitcher plants, that will remain ready to be distributed in case of the extinction of wild populations. This is important for the preservation of all American pitcher plants since the threat of extinction, even for the South American species, can change very swiftly. Even though the South American pitcher plants are not currently threatened, since many species are endemic to very small areas, the risk that a single, massive natural disaster — such as a large wild fire — could wipe out the entire wild population of a species, is very real. The purpose of the Millennium Seed Bank is to provide a safeguard against such a disaster and, consequently, it will serve as an important tool in future conservation initiatives.

Similarly, in the United States, the Center for Plant Conservation based at the Missouri Botanical Garden in St. Louis, Missouri, maintains living collections and seed banks of North America's most imperilled native plants through a network of thirty-three participating botanical gardens and institutions. Within the conservation collections are living plant stocks and banks of seed of *S. rubra* ssp. *jonesii*, *S. oreophila* and many other *Sarracenia* species which are used in repopulation initiatives as part of the CPC's ongoing efforts to preserve and restore wild populations. Both the Millennium Seed Bank and the Center for Plant Conservation provide valuable safeguards against extinction in the wild and represent valuable resources for the study of rare plants and their ecology. Information on the Millennium Seed Bank and Center for Plant Protection projects and ways in which they can be supported is available at:

Royal Botanical Gardens of Kew
  http://www.rbgkew.org.uk/
The Center for Plant Conservation
  http://www.centerforplantconservation.org/

One of the most successful conservation projects of all has been directed through the National Council for the Conservation of Plants and Gardens (NCCPG), a British conservation charity that works directly with horticulturists in establishing a vast network of national collections, each of which covers a particular genus of plants. It is the responsibility of the collection holder "to document, develop and preserve" its collection in trust for the future. The selected collections achieve national collection status only with responsible management and continual documentation that is monitored and assisted by the NCCPG. The ambition of this excellent programme is to permanently maintain a wide range of plants in cultivation as an organized conservation resource. The dedication and commitment of the 630 national collection holders, which in turn represents hundreds of thousands of individual, documented plant species and strains, is truly incredible and certainly a bright hope for the survival of the world's flora. Several vast national collections of American pitcher plants have been set up by dedicated horticulturists and are managed through the NCCPG for the purposes of furthering understanding of pitcher plants and the need to conserve their habitats. Several of the pitcher plants collections are open to the public and can be visited on appointment, please see *http://www.nccpg.com/* for details. The combined bank of NCCPG pitcher plant collections represents the most diverse and comprehensive inventory of American pitcher plants in the world and will be a useful tool in future attempts to restock wild populations.

The International Carnivorous Plant Society, the British Carnivorous Plant Society and the Carnivorous Plant UK internet forum also continue to contribute directly to the conservation of the American pitcher plants through the raising and donating of money to conservation orientated causes including habitat protection and restoration projects. The British Carnivorous Plant Society in conjunction with several British carnivorous plant nurseries donated several hundred carnivorous plants and funds to the Bog of Allen Nature Centre in Lullymore, Ireland, in support of the development and maintenance of an outstanding new display of pitcher plants and other insect-eating plants to promote the need for the conservation of peatland and wetland habitats around the world (see *http://www.ipcc.ie/BOAflytraps.html*). All three organizations are maintained by dedicated volunteers who work relentlessly to further public understanding of these special plants and the need for conservation and are a valuable source of up-to-date information for the carnivorous plant enthusiast. For further information, please see:

The International Carnivorous Plant Society
http://www.carnivorousplants.org/

The (British) Carnivorous Plant Society
http://www.thecarnivorousplantsociety.org/

The Carnivorous Plant UK internet forum
www.cpukforum.com

On a final note, it is important for horticulturists and amateur botanists to understand both that they have a role in the conservation of pitcher plants and what that role is. In the past, private collectors and commercial plant nurseries have poached and plundered wild populations of the American pitcher plants specifically for the purpose of retailing those plants through the horticultural trade. Today, at a time of unprecedented human pressure on the environment, this practice is unethical and completely irresponsible. All of the pitcher plants of the Americas can be sustainably propagated in cultivation and sold or distributed for the enjoyment of horticulturists without any need to deplete and thereby threaten wild populations. Horticulturists can support the conservation of American pitcher plants by only purchasing plants that have been propagated in cultivation from responsible and ethical retailers, some of which are identified in the following chapter. The preservation of American pitcher plants in horticulture is an important safeguard against the total extinction of particular species and represents a reservoir of plant stock that could potentially be used to replenish wild populations in the future.

~

No American pitcher plants were deliberately harmed and certainly none were collected during the field studies that preceded the production of this book.

Cultivation and Horticulture

In addition to their importance as natural members of the American flora (Figure 238), the American pitcher plants are interesting and beautiful plants to grow and study in cultivation. The five genera differ profoundly in terms of growing requirements, but all can be grown easily and vigorously providing that certain fundamental requirements are met.

Sarracenia are by far the easiest of the American pitcher plants to grow; they can be reared with very little effort in a greenhouse, on a sunny windowsill or outside in a bog garden, in pots or at the margins of ponds. D'Amato (1998), Schnell (2002a) and Slack (1979, 1986) provide good overviews of the growing requirements of this genus, but I would like to emphasize the importance of sunlight in the cultivation of these plants. The vast majority of unsuccessful attempts to grow *Sarracenia* fail due to inadequate light levels. As documented in this work, *Sarracenia* naturally grow in extremely open habitat and receive several hours of direct sunlight each day during the growing season. In cultivation, all *Sarracenia* require exposure to direct sunlight in order to growth healthily. Due to its extreme tolerance to cold conditions, the easiest *Sarracenia* species to cultivate is *S. purpurea*. When properly potted and placed in a tray of rainwater in full sun, plants of this species will almost always flourish.

Darlingtonia requires growing conditions that are generally similar to *Sarracenia* except that it displays a clear dislike of high temperatures and prefers slight shade (5–15%) rather than direct sunlight. D'Amato (1998), Schnell (2002a) and Slack (1979, 1986) provide detailed coverage of the horticultural requirements of *D. californica*.

Figure 238 (facing page). *Heliamphora glabra*, like all *Heliamphora* species, occur naturally over small ranges. Cultivated populations of this and other vulnerable species provide buffers against the catastrophic loss of natural populations.

Heliamphora are more complicated than the North American Sarraceniaceae to cultivate successfully since most species require cool, moist conditions duplicating the mild climate of the Guiana Highlands. Providing that low temperatures (below 24°C) can be maintained, most *Heliamphora* generally grow quickly and vigorously. One great misconception which plagues the cultivation of these plants is the idea that *Heliamphora* dislike intense sunlight. As discussed in this work, the vast majority of *Heliamphora* species occur in completely open habitat, exposed to the full intensity of the equatorial sunlight and consequently require extremely intense and long-lasting photoperiods in order to develop excellent colouration and form. It is generally difficult to maintain intense light levels as well as cool temperatures in cultivation, and consequently the majority of *Heliamphora* in cultivation appear etiolated. Slack (1979) and Ziemer (1979) provide detailed accounts of the basic growing requirements of this genus.

Catopsis and *Brocchinia* are relatively rare in cultivation even though they are actually rather easy to grow. All three of the carnivorous bromeliads in these two genera can be grown in a variety of conditions. While they will grow well as companions to *Heliamphora,* they are much more tolerant than *Heliamphora* to warmer and drier conditions, and as is true for most bromeliads, they grow healthily on windowsills and in greenhouses. The basic growing requirements are summarized in D'Amato (1998) and Schnell (2002a). Species in both genera require extremely strong light levels, and preferably cool temperatures, to develop natural colouration and form. Like *Heliamphora,* in the wild these plants grow in direct equatorial sunlight and consequently an intense and long photoperiod should be provided in cultivation.

The International Carnivorous Plant Society, the British Carnivorous Plant Society and the Carnivorous Plant UK internet forum offer the most up-to-date and extensive cultivation advice and information. It is important that horticulturists understand their role within the conservation effort so that a comprehensive approach can be made to secure a future for these beautiful American pitcher plants. Of the many responsible carnivorous plant retailers, I am recommending the following on account of their dedication, reliability and responsible trade ethics, and their long-established pioneering efforts to produce and provide nursery-propogated plants. Each of these specialist operations is devoted to the distribution of carnivorous plants for the enjoyment of horticulturists without threatening or damaging wild populations and each contributes directly and indirectly to the conservation of wild populations of these intriguing and important plants.

Selected Responsible Plant Retailers

Hampshire Carnivorous Plants
Ya Mayla,
Allington Lane,
West End,
Southampton,
SO30 3HQ
United Kingdom

Website: *www.hampshire-carnivorous-plants.co.uk*
See also: *www.hantsflytrap.com*
Email: *matthew@msoper.freeserve.co.uk*
Telephone: +44 (0) 2380 473314
Fax: +44 (0) 2380 473314

Matthew Soper, the founder and manager of this nursery, has grown and studied carnivorous plants for over twenty-five years and has professionally distributed pitcher plants of various genera for more than a decade. His extensive field experience across Southeast Asia, Australia and North America identifies him as a leading authority on various genera of carnivorous plants, especially the tropical pitcher plants, *Nepenthes*.

Hampshire Carnivorous Plants offers an extensive selection of over two hundred species and hybrids of carnivorous plants of the genera *Cephalotus, Darlingtonia, Drosera, Heliamphora, Nepenthes, Pinguicula, Sarracenia* and *Utricularia*. This nursery specializes in mail order services to the United Kingdom, European Union and much of the rest of the world (see its webpage for details). Expansive displays of carnivorous plants are housed in the extensive growing facilities which are open to public viewing on appointment during the growing season (Figure 239). All plants sold by this nursery are produced responsibly through tissue culture, propagation of seed or through division and cuttings.

Hampshire Carnivorous Plants annually attends and displays at a multitude of horticultural shows across the United Kingdom (Figure 240) and has received prestigious gold medal awards every year, for the

Figure 239. Part of the *Sarracenia* collection at Hampshire Carnivorous Plants in Southampton, England.

Figure 240. A prize-winning display, including a number of pitcher plant species, by Hampshire Carnivorous Plants.

last seven years at the Royal Horticultural Society Chelsea Flower Show as well as Best in Show awards at The Hampton Court Palace Show in 2004 and BBC Gardeners World Show in 2005. In both 1998 and 2004, Hampshire Carnivorous Plants also received the esteemed Royal Horticultural Society Anthony Huxley Trophy for its outstanding displays. Indeed, the excellent exhibitions of this nursery alone justify visiting the horticultural shows at which Hampshire exhibits.

Hampshire Carnivorous Plants has operated a long-standing *Sarracenia* breeding programme which led to the creation of the excellent cultivar *S.* 'Juthatip Soper' which was honoured in 1998 with a Royal Horticultural Society Award of Merit. A further excellent cultivar has recently been developed and named *S.* 'Daisy Soper' which will be released in the near future.

~

P&J Carnivorous Plants

> The Hayden,
> Brampton Lane,
> Madley
> Hereford
> HR2 9LX
> United Kingdom

> Website: *www.pj-plants.co.uk*
> Email: *pj_gardner@btopenworld.com*
> Telephone: +44 (0) 1981 251659

P&J Carnivorous Plants developed from Marston Exotics, one of the earliest operating and most highly regarded carnivorous plant nurseries. The founder, Adrian Slack, was fascinated by sundews and other native carnivorous plants which he observed as a child in the 1950s in the English countryside. During the 1960s, he began early attempts at hybridising *Sarracenia* and testing the cultivation and propagation of the genus in the British climate. Over the course of the 1970s he established Marston Exotics and began retailing carnivorous plants commercially. Adrian held lectures, presented talks, wrote avidly and helped found the British Carnivorous Plant Society to further public awareness and understanding. After a trip to the United States in 1974, his enthusiasm on seeing *Sarracenia* in their natural habitat encouraged him to expand the breeding programme at his nursery and, over subsequent years, he produced a vast array of hybrids and a multitude of excellent cultivars, many of which remain unsurpassed in beauty even today. Due to Adrian's tragic decline in health, Marston Exotics was sold in 1987 to Paul Gardner, who continues to distribute carnivorous plants of the highest standard and is committed to maintaining the accomplishments of Adrian.

Marston Exotics closed in 1999 but reopened as P&J Carnivorous Plants during the same year and continues to offer a comprehensive range of over one hundred species in the genera *Cephalotus, Darlingtonia, Drosera, Pinguicula, Sarracenia*, and *Utricularia*. This nursery has a focus on *Sarracenia* and offers mature (4–5 year old) specimens of over forty species and forms and more than seventy hybrids (Figure 241). P&J Carnivorous Plants specializes in mail order to the United Kingdom and European Union, and holds multiple open days each year and welcomes visitors by appointment throughout the growing season. This nursery displays at many horticultural shows across the United Kingdom. All plants sold by this nursery are produced sustainably and responsibly through

Figure 241. Part of the P&J Carnivorous Plants National Collection.

Figure 242. The leaves of *Sarracenia* 'Lynda Butt,' one of Adrian Slack's original cultivars.

the division of stock plants, tissue culture or by growing from seed. P&J Carnivorous Plants directly supports the conservation efforts of the NCCPG through annual donations and regularly holds public talks to promote the preservation and management of carnivorous plant habitats.

Adrian Slack's original *Sarracenia* breeding programme is preserved as the heart of the extensive and well maintained P&J Carnivorous Plants' NCCPG National *Sarracenia* Collection. This collection is a valuable conservation tool, not only as a vast bank of rare *Sarracenia* for potential use in repopulation initiatives but also as a record of horticultural knowledge and experience accumulated over many decades. Adrian's original *Sarracenia* cultivars are conserved and distributed (Figure 242), the very best of which include *S.* 'Burgundy,' *S.* 'Claret,' *S.* 'Daniel Rudd,' *S.* 'Evendine,' *S.* 'Judy,' *S.* 'Loch Ness,' *S.* 'Lynda Butt,' *S.* 'Marston Clone,' *S.* 'Marston Dwarf,' *S.* 'Marston Mill' and *S.* 'Maxima.'

~

Sarracenia Nurseries
37 Stanley Park Road
Carshalton
Surrey
SM5 3HT
United Kingdom

Website: *www.sarracenia.co.uk*
Email: *sarracenia@a4u.com*
Telephone: +44 (0) 2086 477706
Fax: +44 (0) 2086 472259

Sarracenia Nurseries is a longstanding carnivorous-plant retailer that offers a comprehensive selection of the more popular genera, namely *Cephalotus, Darlingtonia, Dionaea, Drosera, Sarracenia* and *Utricularia* as well as an expanding range of *Nepenthes*. This nursery has operated under current management for more than a decade, and the owner, Chris Crow, has more than fifteen years of experience in growing and distributing carnivorous plants of all types. The extensive collections at Sarracenia Nurseries consist of more than 200,000 individual plants and is open for the benefit and education of the public by appointment. A large and permanent display of carnivorous plants in natural-looking conditions is being planned and will soon be constructed to provide visitors with a detailed insight into the dynamics of carnivorous plant habitats and the issues facing their preservation (Figure 243). Sale of plants is conducted primarily through the many British horticultural shows which Sarracenia Nurseries attends and through mail order to customers in the United Kingdom and across the European Union.

Sarracenia Nurseries has directed an extensive *Sarracenia* breeding programme which has concentrated on the production of an original and complete set of *Sarracenia* hybrids. The greatest accomplishment so far is the creation of the outstanding cultivar S. 'Mercury' which produces stocky, 25–35-cm-tall, very dark-coloured pitchers. S. 'Mercury' consists of the parentage S. *(alata x leucophylla) x flava*.

Sarracenia Nurseries is highly involved with the conservation of peat bogs and carnivorous plant wetlands and a generous percentage of the profits of this nursery is donated to these causes. Chris Crow plays an active role in the promotion of sustainably harvested peat as a growing medium for carnivorous plants and, for this reason, Sarracenia Nurseries

Figure 243. Part of the extensive growing facilities at Sarracenia Nurseries in Surry, England.

holds national shares in Irish peat bogs and has donated a wide selection of carnivorous plants to aid the development of a new environmental awareness exhibition centre in Lullymore, Ireland. Carnivorous plants are also frequently donated to the British Carnivorous Plant Society auctions to raise money for the preservation of carnivorous plant habitats around the world. More generally, this nursery adheres to a highly commendable ethos of minimal environmental impact and all power is derived from wind and solar sources and all water requirements are fulfilled through stored rainwater.

Since Sarracenia Nurseries is a nonprofit business operating part-time with a focus on furthering public awareness of conservation issues, Chris Crow regularly works with volunteer groups and welcomes assistance from enthusiasts interested in helping to provide a more secure future for these beautiful plants. Sarracenia Nurseries is easily accessible from London and certainly worth visiting by appointment.

~

Shropshire Sarracenias
5 Field Close
Malinslee
Telford
Shropshire
TF4 2EH
United Kingdom

Website: *www.carnivorousplants.uk.com*
Email: *Mike@carnivorousplants.uk.com*
Telephone: +44 (0) 1952 501598

Shropshire Sarracenias is a specialist operation dedicated to offering an extensive range of rare and popular *Sarracenia* species, hybrids and cultivars. All plants grown at this nursery are arranged and categorized as particular strains so that the identity and history of every plant can be traced and recorded. Currently in excess of one thousand strains of *Sarracenia* are offered, covering virtually all known taxa and discernable variations. All stock is produced through division, so by viewing the established show plants within the nursery collections it is possible to discern the mature appearance of every plant offered for sale. Shropshire Sarracenias operates a mail order service that serves customers world-wide providing that *CITES* channels exist for legal and safe delivery. A smaller range of other carnivorous plant genera is also offered, notably *Darlingtonia* and *Dionaea*.

This nursery opens to public viewing biannually, in midsummer and autumn; see its web link for details. Private viewings of the extensive collection of over 4,500 plants housed in four main greenhouses can also be arranged by appointment during the growing season (Figure 244). Shropshire Sarracenias displays annually at the Shrewsbury flower show and received a silver gilt award in 2002, gold medals in 2003 and 2004 and the Dingle Cup in 2005.

There is an important conservational significance in the manner by which Shropshire Sarracenias operates. This nursery, in effect, forms a sustainable bank of *Sarracenia* plants which may prove a useful resource in future repopulation initiatives. Since a large proportion of the strains reared by this nursery originate from responsibly and ethically collected wild seed, the extensive and accurate records of geographic origin of the stock raises the possibility of repopulating American wetlands

Figure 244. Part of the extensive collection of *Sarracenia* at Shropshire Sarracenias in Shropshire, England.

with plants that originally occurred and belong within the local gene pool. This nursery has been awarded NCCPG national collection status and so multiple specimens of each strain will be permanently maintained and preserved for the benefit of conservation (Figure 8).

~

A comprehensive list of ethical pitcher plant nurseries located in many countries around the world is available at the International Carnivorous Plant Society website; *http://www.carnivorousplants.org/*. Four widely recommended international distributors of pitcher plants include:

United States

California Carnivores
2833 Old Gravenstein Highway South
Sebastopol, California 95472

Telephone: +1 707 824 0433
Fax: +1 707 824 2839
Email: *CALIFCARN@aol.com*
Website: *www.californiacarnivores.com*

Meadowview Biological Research Station
8390 Fredericksburg Turnpike
Woodford, Virginia 22580

Telephone: +1 804 633-4336
Fax: +1 804 633-4336
Email: *meadowview@pitcherplant.org*
Website: *www.pitcherplant.org*

Cooks Carnivorous Plants
PO Box 2594
Eugene, Oregon 97402

Telephone: +1 541 688 9426
Email: *cooks@flytraps.com*
Website: *http://www.flytraps.com*

Australia
Triffid Park
257 Perry Road
Keysborough, Victoria 3173

Telephone: +61 (0)3 9769 1663
Fax: +61 (0)3 9701 5816
Email: *triffids@triffidpark.com.au*
Website: *http://www.triffidpark.com.au/*

~

To purchase pitcher plants from foreign nurseries, it is compulsory to comply with *CITES* import and export regulations and, for most countries, the movement of *CITES Appendix I* species is restricted. See the *Convention on International Trade in Endangered Species* for more information at *www.cites.org*.

Glossary

Ala: Latin for *wing;* refers to the blade-like expansion that runs down the front of the leaves of Sarraceniaceae. Also called keel and wing.

Amazonas: A state of southern Venezuela covered in great part by a vast rainforest and containing several table mountains. Note that the plateaus in Amazonas State are known as cerros, not tepuis.

Annual: A plant that germinates, grows, flowers and reproduces within one year.

Anther: The male part of the flower that produces and contains pollen.

Apical bud: The point from which growth emanates at the apex or terminus of a plant, branch or rhizome. In the context of the American pitcher plants, the apical bud is generally at the centre of the foliage.

Areola: A small pigment-free patch of tissue on the leaves of *Sarracenia* species. Areolation is white and translucent but not transparent as in the fenestration of *D. californica.* Plural: areolae or areoles.

Asexual reproduction: Reproduction involving only one parent and therefore involving no exchange of genes; examples include cuttings, division, production of offshoots, etc.

Bract: A modified leaf-structure present below an individual flower or flower cluster.

Carnivorous: Meat consuming. In the context of this book, carnivorous plants are those which have evolved the ability to trap and digest insects and other small animals.

Cerro: The term for a table mountain in Amazonas State, Venezuela.

CITES: The acronym for *The Convention on International Trade in Endangered Species of Wild Flora and Fauna.*

Complex hybrid: A hybrid that results from the cross-fertilization of two hybrids or one species and one hybrid.

Cotyledon: The first leaves produced after germination.

Crown: See Apical bud.

Cultivar: A particular strain of cultivated plant that exhibits worthy characteristics that are maintained and thereby preserved through the process of cultivation. The rank of cultivar is not the same as the taxonomic category "variety" — the term "cultivar" refers to a specific genetic individual or group exhibiting particular traits.

Cuticle: An impermeable waxy layer present on the foliage of some plants.

Cylindrical: Cylinder shaped.

Dioecious: Single-sexed plants; male and female reproductive organs occur on different individuals of the same species. Reproduction depends upon the presence of two differently sexed plants.

Drainage hole: A small hole present at the front of the midsection of the leaves of most *Heliamphora* species serving to regulate the amount of fluid within the leaf pitcher. Also called a pore**.**

Drainage slit: A short slit present at the front of the pitcher opening of a few *Heliamphora* species.

Ecophene: A form of a plant that exhibits phenotypic characteristics directly a result of the particular habitat in which it occurs.

Endemic: Occurring only in a specific location or region; for example, *Sarracenia* is endemic to North America.

Ensiform: Sword shaped.

Enzyme: A natural catalyst used by organisms in digestion.

Epiphyte: A plant that grows attached to the surface of another plant or plants, usually trees, and generally disconnected from the ground; for example, orchids and bromeliads growing on the branches of tropical trees.

Etiolate: To develop without typical colour due to insufficient amounts of sunlight.

Fenestration: The transparent windows of the leaf of *D. californica*. Note the difference between fenestration and areolation (see Areole).

Flower: The reproductive part of a plant.

Fruit: The swollen seed-bearing pod or case.

Genus: A taxonomic category of organisms consisting of one or more species. For example, *Sarracenia* is a genus of pitcher plants that consists of eight species. All species within a genus will share some, but not all, traits.

Glabrous: Lacking hair.

Gran Sabana: A physiographic region of Bolivar State in southeastern Venezuela; The Gran Sabana is a high plain within the Guiana Highlands.

Growing season: The period of time during which a plant grows. The growing season of both *Darlingtonia* and *Sarracenia* is from spring to early autumn.

Growth point: See Apical bud.

Guiana Highlands: An elevated physiographic region of northern South America, located north of the Amazon River in parts of Brazil, Venezuela, Guyana, Surinam and French Guiana and characterized by numerous isolated plateau-like highlands separated by intervening lowlands.

Hybrid: A crossbreed between different species or different hybrids.

Inflorescence: A cluster of branched flowers borne on a single stem.

Infraspecific taxon: A taxon that is below the rank of a species; for example, a subspecies or a variety.

Infundibular: Trumpet shaped.

Insectivorous: Insect consuming. In the context of this work, insectivorous plants are plants which catch and digest insects.

Intergrade: A transitional form between two taxa. In the context of this work, the term is used mainly to refer to infraspecific hybrids.

Lamina: The blade or expanded part of a leaf.

Lanceolate: Shaped like a lance head; tapering to a point at the apex.

Mesa: Spanish for *table;* a term adopted into English to refer to the great table mountains, or mesas, of the Guiana Highlands.

Monoecious: Double sexed plants; male and female reproductive organs occur on a single individual plant. Most flowering plants are monoecious, and in most cases, reproduction within monoecious plants can take place through self-fertilization.

Ovary: The organ which contains the female sex cells (ovules) that, once fertilized, develops into the fruit and ultimately bears the seed.

Ovate: Egg shaped.

Peduncle: The stalk of a flower.

Perennial: A plant which lives for more than two growing seasons. All of the pitcher plants of the Americas are perennials.

Petal: Often colourful appendages that emanate from the flower for the purpose of attracting pollinators.

Petiole: The leaf stalk. In the case of Sarraceniaceae, the petiole is the non-tubular portion at the very base of the leaf.

Phenotype: The manifestation of visible characteristics of an organism that are determined by the interaction of genetic makeup and environmental influences.

Photosynthesis: The process through which plants produce carbohydrates from water, carbon dioxide and sunlight.

Pistil: The female reproductive portion of a flower.

Pitcher: In the context of this book, the hollow (individual) leaves of Sarraceniaceae.

Pitcher plant: A plant which produces either a hollow leaf or, in the tank bromeliads, a leaf rosette which functions to trap and digest animal prey.

Pollen: The male sex cells which are transferred to the stigma and fertilise the female ovules.

Pubescent: To contain or be covered with short hairs or down.

Reniform: Kidney shaped.

Rhizome: The horizontal, branching stem which develops underground as Sarraceniaceae grow.

Scape: A leafless flower stalk that usually bears a single flower or a tight bunch of flowers.

Sepal: The leaf-like appendages which protect the flower bud as it develops.

Sexual reproduction: Reproduction which involves exchange of genetic material between two organisms; for example; as is required in the production of seeds.

Species: A taxonomic category of organisms that show very similar and closely related traits and are capable of interbreeding.

Stamen: The male reproductive organ found in flowers. Pollen is shed from the anther, a part of the stamen.

Stigma: A part of the pistil which receives pollen. Pollen grains grow through the stigma and fertilise the ovules.

Stolon: A runner or offshoot that grows along or under the ground and eventually grows into a separate plant, but one that is genetically identical to its single parent.

Strain: A particular individual or a specific (genetically identical) lineage descended asexually from a single common ancestor.

Style: The part of the pistil which bears the stigma.

Tank bromeliad: A member of a diverse group of bromeliads which possess modified leaf rosettes that collect and store rainwater and organic debris. In most genera, hollow, water-tight chambers are formed between the leaf axils or at the centre of the leaf rosette and constitute the bromeliads' water-filled tanks. *Brocchinia* and *Catopsis* are tank bromeliads.

Taxon: A taxonomic category at any level. Plural: taxa.

Tepal: A term used instead of petal when the sections of the flower are poorly defined. In the case of the flowers of *Heliamphora*, the petal-shaped structures are technically tepals.

Tepui: A table mountain of the Guiana Highlands, South America. The term "tepui" (also spelled "tepuy") is derived from the Pemón Amerindian word *tepu'u*, meaning "mountain."

Ventricose: Swollen or inflated in shape.

Appendix: Conversion Tables

APPENDIX TABLE 1. LINEAR AND VOLUMETRIC EQUIVALENTS

METRIC MEASUREMENTS		IMPERIAL MEASUREMENTS
1 millimeter [mm]	=	0.03937 in
1 centimeter [cm] (= 10 mm)	=	0.3937 in
1 meter [m] (= 100 cm)	=	1.0936 yd
1 kilometer [km] (= 1000 m)	=	0.6214 mile

IMPERIAL MEASUREMENTS			METRIC MEASUREMENTS
1 inch [in]		=	2.54 cm
1 foot [ft]	(= 12 in)	=	0.3048 m
1 yard [yd]	(= 3 ft)	=	0.9144 m
1 mile	(= 1760 yd)	=	1.6093 km

APPENDIX TABLE 2. THERMAL EQUIVALENTS AND CONVERSIONS[1]

MELTING POINT OF WATER	A WARM DAY	BOILING POINT OF WATER
0° Celsius	26° Celsius	100° Celsius
32° Fahrenheit	78.8° Fahrenheit	212° Fahrenheit

[1]To convert from degrees Celsius to Fahrenheit, multiply the Celsius value by 9/5 and add 32.

To convert from degrees Fahrenheit to Celsius, subtract 32 from the Fahrenheit value and multiply by 5/9.

Bibliography

Baker, J. G. 1882. "New Bromeliads from British Guiana." *Journal of Botany, British and Foreign* 20: 331.

Bell, C. 1949. "A cytotaxonomic study of the Sarraceniaceae of North America." *Journal of the Elisha Mitchell Scientific Society* 65:137–166.

Bentham, G., 1840. "*Heliamphora nutans.*" *Transactions of the Linnean Society of London* 18: 429–432.

Carow, T., A. Wistuba and P. Harbarth. 2005. "*Heliamphora sarracenioides*, a new species of *Heliamphora* (Sarraceniaceae) from Venezuela." *Carnivorous Plant Newsletter* 34(1): 4–6.

Case, F. W., and R. B. Case. 1974. "*Sarracenia alabamensis*, a newly recognized species from central Alabama." *Rhodora* 76: 650–665.

Catesby, M. 1754. *The Natural History of Carolina, Florida, and the Bahama Islands: Volume 1.* London, England: Benjamin White Press.

Clark, C. 1997. *Nepenthes of Borneo.* Kota Kinabalu, Malaysia: Natural History Publications.

———. 2001. *Nepenthes of Sumatra and Peninsular Malaysia.* Kota Kinabalu, Malaysia: Natural History Publications.

Clementi, C. 1916. "A journey to the summit of Mount Roraima." *The Geographic Journal* 48(6): 456–474.

Clusius, C. 1601. *Rariorum Plantarum Historia*, Antwerp, Holland: Plantin-Morens.

Crampton, H. E. 1920. "Kaieteur and Roraima: The great falls and the great mountain of the Guianas." *National Geographic* 38 (3: September): 227–244.

D'Amato, P. 1998. *The Savage Garden: Cultivating Carnivorous Plants.* Berkeley, CA: Ten Speed Press.

Darwin, C. 1875. *Insectivorous Plants.* London, England: John Murray Press.

Determann, R., and M. Groves. 1993. "A new cultivar of *Sarracenia leucophylla* Raf." *Carnivorous Plant Newsletter* 22 (4): 108–109.

Fenner, C. A. 1904. "Beitrag zur Kenntis der Anatomie, Entwickelungsgeschichte und Biologie der Laubblatter und Drusen einiger Insectivoren." *Flora* 93: 335–434.

Fernald, M. 1922. "Notes on the flora of western Nova Scotia." *Rhodora* 24: 165-183.

George, U. 1988. *Inseln in der Zeit: Venezuela Expeditionen zu den letzten weißen Flecken der Erde.* Hamburg, Germany: GEO Verlag.

———. 1989. "Venezuela's islands in time." *National Geographic* 175 (5): 526–561.

Gleason, H. A. 1931. "Botanical results of the Tyler-Duida Expedition." *Bulletin of the Torrey Botanical Club* 58 (6): 367–368.

Gleason, H. A., and E. P. Killip. 1939. "The flora of Mount Auyan-Tepui, Venezuela." *Brittonia* 3: 141-204.

Groves, M. 1993. "Horticulture, trade and conservation of the genus *Sarracenia* in the southeastern states of America." Proceedings of a meeting held at the Atlanta Botanical Garden, September 22-23, 1993.

Hanrahan, R., and J. Miller. 1998. "History of discovery: Yellow flowered *Sarracenia purpurea* L. subsp. *venosa* (Raf.) Wherry var. *burkii*." *Carnivorous Plant Newsletter* 27(1): 14–17.

Im Thurn, E. 1885. "The ascent of Mount Roraima." *Proceedings of the Royal Geographical Society* (7): 497–521.

Lindquist, J. A. 1975. "Bacteriological and ecological observations on the northern pitcher plant, *Sarracenia purpurea*." Masters Thesis, Department of Bacteriology, University of Wisconsin-Madison.

Linnaeus, C. 1753. *Species Plantarum* 1: 510. Sweden, Stockholm: Laurentii Salvii.

Lloyd, F. E. 1942. *The Carnivorous Plants.* New York: Ronald Press.

L'Obel, M. 1581. *Kruydtboeck oft Beschrijvinghe van allerleye Ghewassen, Kruyderen, Hesteren ende Gheboomten.* Antwerp, Holland: Christoffel Plantin.

L'Obel, M., and P. Pena. 1570. *Nova Stirpium Adversaria.* London, England: Purfoot Press.

MacBride, J. 1815. "On the power of *Sarracenia adunca* to entrap insects." *Philosophical Transactions of the Royal Society.* 12: 48-52.

Maguire, B. 1978. "Sarraceniaceae (*Heliamphora*)." *The Botany of the Guyana Highland Part-X, Memoirs of the New York Botanical Garden* 29: 36–61.

Marett, R. R., (ed.), 1934. *Thoughts, Talks and Tramps: A Collection of Papers by Sir Everard Im Thurn.* London, England: Oxford University Press.

Masters, M. T. 1881a. *Gardeners' Chronicle, 2nd Series,* 15: 628–629, 817–818.

———. 1881b. *Gardeners' Chronicle, 2nd Series,* 16: 11–12, 40–41.

Mazrimas, J. A. 1979. "Recent status of *Heliamphora*." *Carnivorous Plant Newsletter* 8 (3): 82–85.

Mellichamp, L. 1979a. "Botanical history of CP IV." *Carnivorous Plant Newsletter* 8 (3): 86–89.

———. 1979b. "The correct common name for *Heliamphora*." *Carnivorous Plant Newsletter* 8 (3): 89.

Meyers-Rice, B. 1997. "An anthocyanin-free variant of *Darlingtonia californica*: Newly discovered and already imperiled." *Carnivorous Plant Newsletter* 26 (4): 129–132.

———. 1998. "*Darlingtonia californica* 'Othello'." *Carnivorous Plant Newsletter* 27 (2): 40–42.

Mez, C. C. 1896. *Monographiae Phanerogamarum* 9: 621.

———. 1913. *Repertorium Specierum Novarum Regni Vegetabilis* 12: 414.

Michaux, A. 1803. *Flora Boreali-Americana* 1: 311.

Nerz, J. 2004. "*Heliamphora elongata* (Sarraceniaceae), a new species from Ilu-Tepui." *Carnivorous Plant Newsletter* 33(4): 111–116.

Nerz, J., and A. Wistuba. 2000. "*Heliamphora hispida* (Sarraceniaceae), a new species from Cerro Neblina, Brazil-Venezuela." *Carnivorous Plant Newsletter* 29(2): 37–41.

———. 2006. "*Heliamphora exappendiculata*, a clearly distinct species with unique characteristics." *Carnivorous Plant Newsletter* 35(2): 43–51.

Nerz, J., A. Wistuba, and G. Hoogenstrijd. 2006. "*Heliamphora glabra* (Sarraceniaceae), eine eindrucksvolle *Heliamphora* Art aus dem westlichen Teil des Guayana Schildes." *Das Taublatt* 54: 58–70.

Pietropaolo, J., and P. Pietropaolo. 1986. *Carnivorous Plants of the World*. Portland, OR: Timber Press.

Rafinesque, C. S. 1817. *Florula Ludoviciana; or A flora of the State of Louisiana. Translated, revised, and improved from the French of C. C. Robin*. New York, NY: C. Wiley & Co.

Robbins, C. S. 1998. "Examination of the U.S. Pitcher-plant Trade With a Focus on the White-topped Pitcher-plant" *Traffic Bulletin*. 17(2) 79-86.

St. John, S. 1862. *Life in the Forests of the Far East. Volume 1*. London, England: Smith & Elder Press.

Schnell, D. E. 1978. "*Sarracenia rubra* Walter: Infraspecific nomenclatural corrections." *Castanea* 43: 260–261.

———. 1979. "*Sarracenia rubra* Walter subspecies *gulfensis*: A new subspecies." *Castanea* 44: 217–223.

———. 1993. "*Sarracenia purpurea* L. subsp. *venosa* (Raf.) Wherry var. *burkii* Schnell (Sarraceniaceae): A new variety of the gulf coastal plain." *Rhodora* 95: 6–10.

———. 1998. "*Sarracenia flava* L. varieties." *Carnivorous Plant Newsletter* 27(4): 116–120.

———. 2002a. *Carnivorous Plants of the United States and Canada*. Portland. OR: Timber Press.

———. 2002b. "*Sarracenia minor* Walt. var. *okefenokeensis* Schnell: A New Variety." *Carnivorous Plant Newsletter* 31(2): 36–39.

Schnell, D. E., and R. Determann. 1997. "*Sarracenia purpurea* L. ssp. *venosa* (Raf.) Wherry var. *montana* Schnell & Determann (Sarraceniaceae): A new variety." *Castanea* 62: 60–62.

Schomburgk, R. H. 1840. "Journey from Fort San Joaquim, on the Rio Branco, to Roraima, and thence by the rivers Parima and Merewari to Esmeralda, on the Orinoco, in 1838–9." *Journal of the Royal Geographical Society* 10: 191–247.

———. 1849. *Reisen in Britisch Guiana in den Jahren 1840–1844*. Leipzig, Germany: Im Aufrag Sr. Majestät des Königs von Preussen.

Slack, A. 1979. *Carnivorous Plants*. London, England: Ebury Press.

———. 1986. *Insect-Eating Plants and How to Grow Them*. London, England: Alpha Books.

Steyermark, J. 1951. "Sarraceniaceae." *Fieldiana, Botany* 28: 239–242.

———. 1984. "Flora of the Venezuelan Guayana." *Annals of the Missouri Botanical Garden* 71: 297–340.

Tate, G. H. H. 1930. "Through Brazil to the summit of Mount Roraima." *National Geographic* 58 (5: November): 584–605.

Taylor, D. 1998. *South Carolina Naturalists: An Anthology, 1700–1860.* Columbia, SC: University of South Carolina Press.

Torrey, J. 1853. "On *Darlingtonia californica*, a new pitcher plant from northern California." *Smithsonian Contributions to Knowledge* 6(4): 5.

Walter, T. 1788. *Flora Caroliniana, secundum.* London, England: J. Fraser.

Wherry, E. 1933. "The geographic relations of *Sarracenia purpurea.*" *Bartonia* 15: 1–8.

———. 1972. "Notes on *Sarracenia* subspecies." *Castanea* 37: 146–147.

Whitely, H. 1884. "Explorations in the neighbourhood of Mount Roraima and Kukenam in British Guiana" *Proceedings of the Royal Geographical Society* 7: 452–464.

Wilson, P. 2001. "*Sarracenia leucophylla* 'Schnell's Ghost'." *Carnivorous Plant Newsletter* 30 (1): 11–14.

Wistuba, A., P. Harbarth, and T. Carow. 2001. "*Heliamphora folliculata*, a new species of *Heliamphora* (Sarraceniaceae) from the 'Los Testigos' Table Mountains in the south of Venezuela." *Carnivorous Plant Newsletter* 30(4): 120–125.

Wistuba, A., T. Carow, and P. Harbarth. 2002. "*Heliamphora chimantensis*, a new species of *Heliamphora* (Sarraceniaceae) from the 'Macizo de Chimanta' in the south of Venezuela." *Carnivorous Plant Newsletter* 31(3): 78–82.

Wistuba, A., T. Carow, P. Harbarth and J. Nerz. 2005. "*Heliamphora pulchella*, eine neue mit *Heliamphora minor* (Sarraceniaceae) verwandte Art aus der Chimante Region in Venezuela." *Das Taublatt* 53: 3.

Wood, A. 1863. *Leaves and Flowers, or Object Lessons in Botany, with a flora.* New York, NY: A. S. Barnes & Company.

Ziemer, R. 1979. "Some personal observations on cultivating the *Heliamphora.*" *Carnivorous Plant Newsletter* 8 (3): 90–92.

About the Author

From an early age, travel as a window into different countries and landscapes revealed to me the extraordinary diversity and beauty of the natural world. From my first experiences abroad, traveling with my family, I vividly remember being overwhelmed and fascinated by the vibrant array of plants and animals that surrounded me. From the age of seven I assembled a diverse collection of plants, insects and other animals which I could observe and study whilst at my home in the south of England, and in the subsequent years my fascination for the natural world continued to grow.

During my late teenage years I took part in various rainforest conservation programmes in several countries in Central America and Southeast Asia and traveled as frequently as possible to equatorial countries to observe tropical wildlife first hand (Figure 245). To further my understanding of the wider world, I studied Geography at the University of Durham, England, and incorporated into my degree a year at the University of Tuebingen in Germany and an additional year at Yale University in the United States to better understand the international scope of the discipline.

I first encountered the American pitcher plants at the age of eight and I remember being enthralled by the extraordinary shape and intriguing carnivorous nature of the plants' strange tubular leaves. During the thirteen years since, I have cultivated and studied pitcher plants from all five genera and yet still this remarkable group of plants continues to amaze me. In response to the lack of detailed information regarding several genera of carnivorous plants, I began writing *Pitcher Plants of the Americas* at the age of sixteen with the intent of completing it before turning twenty-two years old. After five years of research and field trips across the American continents, this project is now complete. It has been a challenging, but at the same time a profoundly rewarding, experience and I sincerely hope you will enjoy the results.

Figure 245. The author in Borneo while researching types of prey caught by members of the Old World pitcher plant genus *Nepenthes*.

While researching *Pitcher Plants of the Americas*, I traveled to the Guiana Highlands of South America and became fascinated with both the remarkable diversity of life on the summits of the tepuis and the incredible stories that were part of the discovery and exploration of the region by Euro-Americans. As a result, for the past two years, I have been investigating the unique natural history of these dramatic and ancient tablelands in preparation for my second book, *Lost Worlds*.

I also continue to explore the adaptations of organisms to survival in extreme conditions, with particular attention to the carnivorous plants. At this time, I am focusing my efforts on the pitcher plants of the Old World and the more cosmopolitan sticky "fly-paper" plants, but in general I find myself exploring the entire realm of carnivorous plants — particularly their natural history, biogeography, uses in horticulture, and conservation status — and contemplating the ways that I might be able to make a contribution to their protection and future well-being.

Stewart McPherson

www.redfernnaturalhistory.com

Image Credits

Several images included in this book were provided by other individuals and organizations, and I would like to sincerely thank all of these donors for their generosity in helping to make this project better than it otherwise would have been. The donors are listed below in alphabetical order, and the numbers that follow their names identify the figures for which credit is given. All images that are not credited here were taken by the author with the exception of some *Heliamphora* images that were provided anonymously.

The CAMENA Project, Heidelberg University and Mannheim University: 147
Chris Crow: 243
Paul Gardner: 241, 242
Gert Highbattle: 108
Debbie M. Jolliff, *www.debbiejolliff.co.uk*: 240
Mike King: 8, 244
Jerry N. McDonald: 7, 234
Jim Miller: 220
Dr. Barry Rice: 206, 207, 208, 216, 217, 231
Fernando Rivadavia: 140
Andy Smith: 24, 25, 38, 39, 47, 50, 72, 73, 96, 139, 151, 152
Matthew Soper: 239
The University of Wisconsin-Madison: 148, 149

Index

Index